Interferometry of Fibrous Materials

The Adam Hilger Series on Optics and Optoelectronics

Series Editors: **E R Pike** FRS and **W T Welford** FRS

The Adam Hilger Series on Optics and Optoelectronics

Interferometry of Fibrous Materials

N Barakat CPhys FInstP

Ain Shams University, Egypt

and

A A Hamza CPhys FInstP FRMS

Mansoura University, Egypt

Adam Hilger, Bristol and New York

British Library Cataloguing in Publication Data

Baraket, N.
   Interferometry of fibrous materials.
   1. Fibres. Analysis. Applications of interferometry
   I. Title   II. Hamza, A. A.
   677

   ISBN 0-85274-100-6

Library of Congress Cataloging-in-Publication Data

Barakat, N., 1922–
    Interferometry of fibrous materials / by N. Barakat, A. A. Hamza.
      p. cm.—(The Adam Hilger series on optics and
optoelectronics)
    ISBN 0-85274-100-6
    1. Fibers—Optical properties. 2. Interferometry. I. Hamza, A.
A. II. Title. III. Series.
TA418.9.F5B35 1990
620.1′9704295—dc20   89-19960
                                             CIP

The authors have attempted to trace the copyright holder of all the figures and tables reproduced in this publication and apologise to copyright holders if permission to publish in this form has not been obtained.

Published under the Adam Hilger imprint by IOP Publishing Ltd
Techno House, Redcliffe Way, Bristol BS1 6NX, England
335 East 45th Street, New York, NY 10017-3483, USA

Typeset by Mathematical Composition Setters Ltd, Salisbury, England
Printed in Great Britain by J W Arrowsmith Ltd, Bristol

To:
Afaf, Hani, Hala and Heba Barakat
Sahar, May, Sarah and Roaa Hamza

Contents

Series Editors' Preface

Optics has been a major field of pure and applied physics since the mid 1960s. Lasers have transformed the work of, for example, spectroscopists, metrologists, communication engineers and instrument designers in addition to leading to many detailed developments in the quantum theory of light. Computers have revolutionised the subject of optical design and at the same time new requirements such as laser scanners, very large telescopes and diffractive optical systems have stimulated developments in aberration theory. The increasing use of what were previously not very familiar regions of the spectrum, e.g. the thermal infrared band, has led to the development of new optical materials as well as new optical designs. New detectors have led to better methods of extracting the information from the available signals. These are only some of the reasons for having an *Adam Hilger Series on Optics and Optoelectronics*.

The name Adam Hilger, in fact, is that of one of the most famous precision optical instrument companies in the UK; the company existed as a separate entity until the mid 1940s. As an optical instrument firm Adam Hilger had always published books on optics, perhaps the most notable being Frank Twyman's *Prism and Lens Making*.

Since the purchase of the book publishing company by The Institute of Physics in 1976 their list has been expanded into all areas of physics and related subjects. Books on optics and quantum optics have continued to comprise a significant part of Adam Hilger's output, however, and the present series has some twenty titles in print or to be published shortly. These constitute an essential library for all who work in the optical field.

Preface

Interferometry has made important contributions in a variety of different fields in physics, applied physics and engineering. It has acquired importance by making available sensitive tools for a wide variety of physical measurements, including controlling production in the lens, surface finish, metrology and fibre industries.

There exists a real need for an up-to-date text on interferometry applied to fibres. Indeed a graduate may find at his or her disposal a book on fibre interferometry which will minimise the need to refer to original papers. But for those who might wish to pursue the subject further, numerous references to the literature are provided.

A brief comment as to the contents. The book describes the use of interferometric techniques for fibre analysis and research. Two-beam and multiple-beam interferometric methods are given, together with their application in the investigation of the properties of fibres. Attention is paid to basic ideas leading to the formation of interferograms. Deduction of data from interferograms by digitising the images and extraction of index profiles of fibres by the aid of computer programs are presented.

The book includes appreciable material resulting from research activities on the applications of interferometry to fibres. Emphasis has been directed to the development and use of the many different kinds of interference microscopes now manufactured on an international scale.

Our aim in compiling the contents of the book is to achieve the following.

(*a*) To present the phenomenon of the interference of light, to review interferometers and their application to fibres and to discuss the different but related aspects of the wide field of applications.

(*b*) To keep the mathematical level of treatment as low as possible so as to avoid putting a double load on scientists or engineers working in the fibre industry who have to assimilate a large volume of knowledge.

(*c*) To assist scientists and engineers working in the field of fibre interferometry, either for higher studies or in industry, to have a wide grasp of the field and its applications.

We acknowledge the many useful comments and general suggestions for improvement that were provided by Professor W T Welford and Professor E R Pike, to whom we are very grateful.

N Barakat and A A Hamza
Cairo, 1989

Acknowledgments

We are grateful to the following for making various diagrams and tables available to us and for granting permission to reproduce them in this work.

IOP Publishing Ltd for figures 1.1, 1.2, 4.8, 4.9, 4.10, 4.11, 4.19 and 4.20, and table 1.2.

McGraw-Hill Publishing Company for figure 1.5.

Textile Research Institute for figures 2.9, 4.3, 4.5 and 4.14.

Journal of Microscopy for figures 3.4, 3.12, 7.6 and 7.7.

Optical Society of America for figures 4.13, 5.7, 5.8 and 6.1.

Elsevier Applied Science Publishers Ltd for figure 4.21.

The Textile Institute for figure 5.5.

Taylor & Francis Ltd for figure 6.2.

Pergamon Press PLC for figures 6.3 and 6.4.

Longman for figures 7.2 and 7.3.

Carl Zeiss Jena for figure 7.5.

1 An Introduction to Fibre Structure

The present chapter comprises the following topics: methods of investigating the structure of fibres; optical anisotropy in fibres (natural and synthetic fibres and highly oriented fibres); layer structure in synthetic fibres; optical fibres.

1.1 Methods of investigating the structure of fibres

The various methods are listed below:

(a) optical microscopy;
(b) scanning electron microscopy;
(c) transmission electron microscopy;
(d) x-ray spectrometry;
(e) infrared spectroscopy;
(f) two-beam interference microscopy; and
(g) multiple-beam interference in transmission and at reflection.

For optical and electron microscopy the reader is advised to consult references in the field, for example Stoves (1957), Françon (1961), Meredith and Hearle (1959) and Wells (1974). The scanning electron microscope, due to its high spatial resolution and great depth of field, can provide detailed information on the structural features of fibres. For optical fibres, profile information could be obtained by etching the end face of the fibre and making use of the fact that the etch rate is dependent on the local composition.

When scanning electron microscopy is equipped with an energy dispersive x-ray spectrometer, quantitative information about the effect of variations in composition may be obtained (Wells 1974, Kita *et al* 1971). By comparing the characteristics of x-ray intensities of the fibre with those of a standard sample, the composition of the fibre can be determined for elements not higher than Be (Burrus *et al* 1973).

Chapters 3 and 7 deal with the theory and formation of two-beam interference systems and the variants of interference microscopes and their application to the determination of the optical properties of textile and optical fibres.

Chapter 4 is concerned with the theory, formation and characteristics of multiple-beam interference fringes in transmission and at reflection and their application in the determination of the optical properties of natural, synthetic and optical fibres.

1.2   Optical anisotropy in fibres

1.2.1   Natural and synthetic fibres

When a plane polarised beam of monochromatic light passes through a system of oriented molecules it suffers refraction, and the refractive index is determined by the interaction of light with matter. This interaction differs with the direction of the electric vector of the incident plane polarised light. The two directions of the electric vector are: (i) along the molecular axis and (ii) perpendicular to it. The material is termed optically anisotropic and gives rise to double refraction (birefringence) which is the difference between the two refractive indices for the light vibrations in the parallel and perpendicular directions.

Natural and synthetic fibres consist of long-chain molecules. These molecules lie along the fibre axis; in some fibres they are almost exactly parallel and in others they are only roughly parallel. Optical properties of fibres vary with direction in the fibre and the maximum difference is between the properties along the fibre axis and those perpendicular to it. There is a direct relation between the optical properties of fibres and those of the individual molecules constituting these fibres, because there is little interaction between the neighbouring molecules within the fibre. The refractive index of the fibre material for any particular direction of light vibration is approximately the sum of the properties of individual molecules constituting the fibre in this direction (see Bunn 1949). For studying the optical properties of fibres, plane polarised light vibrating parallel and perpendicular to the fibre axis is used. There are many methods for determining the refractive indices of fibres, amongst which are the Becke-line method and interferometric methods.

The Becke-line method gives the refractive index of the outer layer of the fibre, which may be different structurally from the interior (Hartshorne and Stuart 1970). In the Becke-line method, a line of light appears at the boundary between the fibre and an immersion liquid of known refractive index. When observed through a microscope, this line moves towards the medium of higher refractive index when the microscope objective is raised. Using a set of standard liquids of known refractive indices, it is possible to investigate different fibres of widely varying refractive indices. When the refractive index of the liquid matches that of the fibre, the line of demarcation between each medium disappears. This is termed the matching liquid, and accordingly the refractive index of the fibre is determined for the vibration direction of the plane polarised light illuminating the fibre and of this wavelength. Birefringence Δn (double refraction) of a fibre is the difference between the refractive index $n^{\parallel}$ of the fibre along its axis and $n^{\perp}$ across the fibre axis, $\Delta n = n^{\parallel} - n^{\perp}$.

Application of two-beam and multiple-beam interferometric methods presented in Chapters 3 and 4 leads to the determination of refractive indices of the skin and core of fibres and their birefringence Δn_s and Δn_c for the two directions of light vibrations, parallel and perpendicular to the fibre axis.

The first theory to yield a relation between the molecular structure of a uniaxially oriented polymer and its optical anisotropy was developed by Kuhn and Grün (1942). Synthetic fibres spun from the melt of a certain polymer by extrusion through fine holes are almost isotropic in their physical properties. To produce a stronger fibre from these spun fibres, suitable for industrial use, they are mechanically drawn. The fibre in the drawn or extended state shows considerable optical and mechanical anisotropy. The degree of anisotropy may be dependent on the amount of extension imposed. The values of the mean refractive indices $n_a^{\parallel}$ and $n_a^{\perp}$ and the mean birefringence Δn_a give valuable information for the characterisation of these fibres.

Refractive indices and bond polarisability
Assessment of the optical anisotropy of fibres is of considerable importance, since it provides information about the degree of orientation of a particular molecular system, making it possible to evaluate the effect of any chemical or mechanical treatments; knowledge of this kind is of great relevance in the development of modern methods of quality control in many industrial processes. Denbigh (1940) presented a method of assessing the molecular anisotropy based on the concept of bond polarisability and an empirical scheme of individual polarisabilities for all types of chemical bonds encountered in a given structure. In his treatment each different type of chemical bond is associated with its own polarisability and the contributions of all bonds are added. This would allow calculation of the refractive index of the given structure. The average polarisability for each bond cannot be added in the case of directional properties in anisotropic crystals. It is necessary in this case to use ellipsoid polarisability. The following equation, given by Bunn (1961), gives the polarisability α in a

principal direction of the polarisability ellipsoid of a polyatomic molecule

$$\alpha = \Sigma \; \alpha_L \cos^2 \theta + \Sigma \; \alpha_T \sin^2 \theta$$

summed for all bonds, where θ is the angle between a bond and the direction in question and α_L and α_T are longitudinal (along the bond) and transverse (across the bond) polarisabilities respectively.

The same expression applies to the case of a crystal where the polarisability for a principal direction of the ellipsoid can be calculated. The refractive index n for that direction is calculated from the polarisability using the Lorentz–Lorenz expression

$$\frac{n^2 - 1}{n^2 + 2} \frac{M}{d} = \frac{4}{3} \pi N \alpha_k$$

where n is the appropriate ($n^{\parallel}$ or $n^{\perp}$) refractive index, M is the molecular weight per repeat length, d is the density of the substance, N is Avogadro's number and α_k is the index of polarisability of the whole repeat unit of the polymer chain.

Hamza and Sikorski (1978) calculated the refractive indices and birefringence of poly(p-phenylene terephthalamide) (PPT) fibres, based on a molecular model of the PPT polymer given by Northolt (1974), together with the values of bond polarisabilities given by Denbigh (1940) and by Bunn and Daubeny (1954). The Lorentz–Lorenz expression was used.

1.2.2 *Highly oriented fibres*
The PPT polymer mentioned above belongs to the groups of highly oriented fibres also termed high-performance synthetic organic fibres. A fibre that possesses a high tensile modulus greater than $40 \; \text{GN} \, \text{m}^{-2}$, can be classified as a high-performance fibre. This figure effectively excludes all the conventional textile fibres, including high-tenacity nylons and polyethylene terephthalate fibres. The studies on polymer crystallisation presented by Keller (1968) which demonstrated the tendency of polymers to form folded-chain crystals, gave a clear understanding of the relationship between fine structure and physical properties. PPT fibres are examples of high-performance organic fibres; Kevlar and Twaron are trade names. Kevlar 49 fibres have outstanding physical properties, for example remarkably high tensile strength. The stress–strain curves of these fibres show a linear relationship indicating elastic behaviour over very small extensions. Since Kevlar 49 has an outstanding specific strength (ratio of the strength to specific gravity), it is useful as a reinforcement for composite materials.

Carter and Schenk (1975) used x-ray diffraction and birefringence measurements to correlate the physical properties of high-performance fibres with their structure. There exists a close correlation between the high modulus of these fibres and their preferred orientation. High-angle x-ray diffraction was used to study the high degree of lattice order. Northolt

(1974) evaluated the unit cell dimensions of PPT fibres as monoclinic with $a = 0.719$ nm, $b = 0.518$ nm, $c = 1.29$ nm and $\gamma = 90°$. A detailed review of the structure and physical properties of PPT fibres is given by Dobb and McIntyre (1984).

The refractive indices and birefringence of fibres $n^{\parallel}$ and $n^{\perp}$ provide a convenient measure of the extent of molecular alignment along the fibre axis and perpendicular to it. Measurement of the birefringence provides a measure of the degree of molecular orientation and closeness of molecular packing at different regions of anisotropic fibres. Such interferometric measurements are often made in conjunction with investigations by x-ray diffraction, electron microscopy and infrared spectroscopy.

Hamza and Sikorski (1978) calculated the principal refractive indices and birefringence of PPT fibres and applied two-beam interference using the Pluta microscope with monochromatic and white light, vibrating parallel and perpendicular to the fibre axis (see Chapters 3 and 7). Kevlar 49 fibre was immersed in a liquid of $n_L = 1.656$. The following values were reported for the refractive indices $n^{\parallel}$, $n^{\perp}$ and birefringence Δn of Kevlar: $n^{\parallel} = 2.267$, $n^{\perp} = 1.605$ and $\Delta n = 0.662$, for $\lambda = 546$ nm. Notice the remarkably high value of the birefringence. Hamza and Sikorski also discussed and compared the theoretical and experimentally determined refractive indices.

Table 1.1 gives the values of refractive indices n and the birefringence Δn of some natural and synthetic fibres.

Table 1.1 Refractive indices $n^{\parallel}$ and $n^{\perp}$ and birefringences Δn of some natural and synthetic fibres.

Fibre	$n^{\parallel}$	$n^{\perp}$	Δn	Reference
Cotton	1.578	1.532	0.046	Preston (1933)
Ramie and flax	1.596	1.528	0.068	Preston (1933)
Viscose rayon	1.539	1.519	0.020	Preston (1933)
Viscose rayon (skin)	1.5563	1.5282	0.0281	Faust (1952)
(core)	1.5536	1.5304	0.0234	Faust (1952)
Viscose rayon (skin)	1.5453	1.5226	0.0227⎫	Barakat and Hindeleh
(core)	1.5441	1.5247	0.0194⎭	(1964)
Wool	1.557	1.547	0.010	Hartshorne and Stuart (1970)
Polyethylene	1.574	1.522	0.052	Hartshorne and Stuart (1970)
Polypropylene	1.530	1.496	0.034	Hartshorne and Stuart (1970)
Acrilan	1.517	1.519	− 0.002	Barakat and El-Hennawi (1971)
Acrilan	1.511	1.514	− 0.003	Hartshorne and Stuart (1970)
Nylon 6	1.575	1.526	0.049	Hartshorne and Stuart (1970)
Nylon 6 (skin)	1.5533	1.5448	0.0085	Hamza et al (1985b)
(core)	1.5512	1.5430	0.0082	Hamza et al (1985b)
Nylon 66	1.578	1.522	0.056	Hartshorne and Stuart (1970)
Terylene	1.706	1.546	0.160	Hartshorne and Stuart (1970)
Dralon	1.5201	1.5234	− 0.0033	Hamza et al (1985b)
Kevlar 49	2.267	1.605	0.662	Hamza and Sikorski (1978)

1.2.3   Layer structure in synthetic fibres

Multiple-beam interference fringes were applied by Hamza and Kabeel (1986) to the measurement of refractive indices and birefringence of polypropylene fibres. The fringe system resolved layer structure in the undrawn fibre, and the refractive indices and the extension of each layer were measured. Figure 1.1 shows microinterferograms of multiple-beam Fizeau fringes in transmission for a polypropylene fibre. Plane polarised light of wavelength 546.1 nm vibrating parallel and perpendicular to the fibre axis illuminated the interferometer successively. The fibre was immersed in a silvered liquid wedge, n_L being 1.5015 at 22.5 °C. From the shape of the fringe shift inside the fibre, three resolved layers are observed. Only the results of the determination of refractive indices and birefringence of each of the three layers as well as their extension are reported here, since the theory of multiple-beam Fizeau fringes applied to fibres of multilayer structure is given in Chapter 4. Table 1.2 gives the refractive indices and birefringence of polypropylene fibre layers using $\lambda = 546.1$ nm at 22.5 °C. The radii of the three layers are $r_1 = 36.3$ μm, $r_2 = 18.8$ μm and $r_3 = 9.6$ μm, $r_1 = r_f$ the fibre radius. As explained in Chapter 4, each layer contributes to the shape of the fringe shift Z across the fibre cross section extension X by

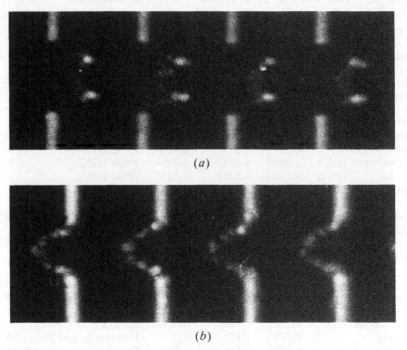

(a)

(b)

Figure 1.1 Multiple-beam fringes crossing a polypropylene fibre revealing layer structure. Monochromatic light vibrating parallel (a) and perpendicular (b) to the fibre axis. (From Hamza and Kabeel 1986.)

Table 1.2 Refractive indices and birefringence of polypropylene fibre layers, when using light of wavelength 546.1 nm and at a temperature of 22.5 °C. (From Hamza and Kabeel 1986.)

Refractive index of liquid†, n_L	The mean refractive indices and mean birefringence of the fibre‡			Refractive indices and birefringence of fibre layers‡								
				First layer (outer layer)			Second layer			Third layer (core)		
	$n_a^\parallel$	$n_a^\perp$	Δn_a	$n_1^\parallel$	$n_1^\perp$	Δn_1	$n_2^\parallel$	$n_2^\perp$	Δn_2	$n_3^\parallel$	$n_3^\perp$	Δn_3
1.5015	1.5028	1.5001	0.0027	1.5032	1.5007	0.0025	1.5015	1.5000	0.0015	1.5014	1.4995	0.0019

† The error in measuring n_L using an Abbe refractometer is ±0.0002.
‡ The error in $n^\parallel$ and $n^\perp$ is ±0.0007.

half an ellipse on the (Z, X) plane where the interference fringes are formed. Figure 1.2 gives the shape of the fringe shift for light vibrating parallel and perpendicular to the fibre axis respectively.

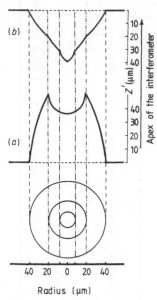

Figure 1.2   Calculated shape of the fringe shift versus distance from the fibre centre for light vibrating parallel (*a*) and perpendicular (*b*) to the fibre axis. (From Hamza and Kabeel 1986.)

1.3   The structure of optical fibres

1.3.1   Types of optical fibres

An optical fibre consists, in its simplest form, of a coaxial arrangement of two homogeneous silica glasses as given in figure 1.3. It is a step-index optical fibre as shown from its refractive index profile. The core material has a refractive index higher than that of the cladding material. Most optical fibres have more than two layers. Figure 1.4 shows an optical fibre consisting of an inhomogeneous core surrounded by a cladding region and a plastic jacket to protect the fibre from scratches and other physical causes of damage. The fibre shown in figure 1.4 is termed a graded-index optical fibre, in which the core refractive index distribution varies as a function of the radial coordinate with maximum refractive index at the centre of the fibre. Figure 1.5 gives the cross section and the refractive index profile of some types of optical fibres and dimensions of the core and cladding. The fibre types are:

(*a*) step-index fibre in which the core is a dielectric rod and the cladding is air;

(*b*) multimode step-index optical fibre;

(*c*) single-mode step-index optical fibre;

(*d*) W-type fibres in which the refractive index profile takes the shape W and the core is surrounded by a double cladding, the inner cladding having refractive index n_1 smaller than the refractive index n_2 of the outer cladding;

(*e*) multimode graded-index optical fibre in which the core refractive index $n(r)$ is related to the radial coordinate r, with its origin at the centre, by the relation

$$n^2(r) = n^2(0)\left[1 - \Delta_1\left(\frac{r}{a}\right)^\alpha\right]$$

where

$$\Delta_1 = \frac{\Delta^2}{n^2(0)} = \frac{n^2(0) - n_1^2}{n^2(0)}$$

and $1.5 < \alpha < 2.5$. The refractive index $n(r)$ attains its maximum value at the fibre axis;

(*f*) graded-index optical fibre of the W-type.

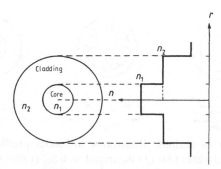

Figure 1.3   Cross section of step-index fibre showing core, cladding and index profile.

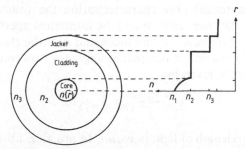

Figure 1.4   Cross section of graded-index fibre showing core, cladding, jacket and index profile.

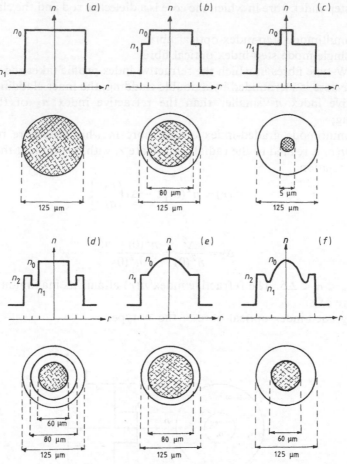

Figure 1.5 Cross sections, dimensions and index profiles of different types of optical fibres (a)–(f) described in text. (From Costa 1980.)

For step-index optical fibre characterisation the following parameters specify the fibre: the core radius a and the numerical aperture (NA) defined by $(n_0^2 - n_1^2)^{1/2}$, where n_0 is the core index and n_1 is the cladding index. The NA is related to the maximum acceptance angle for rays entering the fibre by another parameter V given by

$$V = \frac{2\pi a}{\lambda} (n_0^2 - n_1^2)^{1/2}$$

where λ is the wavelength of light *in vacuo*. In practice, fibres have values of $\Delta \ll 1$, typically $\leqslant 0.2$.

An optical fibre of the multimode type is a dielectric waveguide which has many propagation modes. These modes are periodic field distributions, which, when taken together, can be used to construct any allowed field distribution in the fibre. Figure 1.6 shows a representation of the light paths in these modes by rays. Three regions, 1, 2 and 3, are indicated, representing the core, cladding and coating respectively. Figure 1.6(*a*) represents a single-mode fibre while figure 1.6(*b*) shows a graded-index multimode fibre.

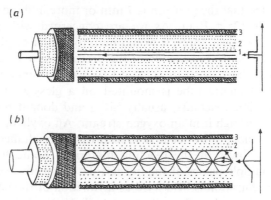

Figure 1.6 Propagation of light waves in (*a*) step-index single-mode and (*b*) graded-index multimode fibre.

1.3.2 Compositional characteristics of optical fibres

Optical fibres used in optical communication are basically of two types: single-mode and graded-index multimode fibres. In a multimode graded-index fibre, the central region, the core, typically consists of doped silica glass. The low refractive index of pure SiO_2, $n_0 = 1.450$ at $\lambda = 1.0\ \mu m$, is modified by doping with materials such as oxides of germanium, phosphorous and boron (Rigterink 1975). Germanuim (MacChesney *et al* 1974) and phosphorous (Payne and Gambling 1974) increase the refractive index of SiO_2, boron decreases it (French *et al* 1973). The fibre can guide light waves if the core has a higher refractive index than the surrounding region (Marcuse 1972). Thus most fibres consist of a cladding region of pure SiO_2 and a core whose index is increased by addition of germanium or phosphorous oxides. In some fibres the cladding is doped with boron oxide to lower its refractive index relative to the undoped fibre core.

The amount of dopant that is added plays an important role, since the more dopant the greater the acceptance angle of the fibre and its resulting numerical aperture. On the other hand, increasing the dopant increases compositional fluctuations which lead to fibre losses by scattering and introduces fabrication difficulties due to mismatches in the physical properties of the core and cladding glasses (Marcuse and Presby 1980). Typical germanium-doped fibres have a maximum index difference of 0.02. Step-index and graded-index optical fibres are fabricated by the modified chemical vapour deposition (MCVD) process (see MacChesney *et al* 1974). Losses as low as $0.2\,dB\,km^{-1}$ at $\lambda = 1.55\,\mu m$ have been achieved (Miya *et al* 1979). The MCVD method comprises two processes: production of a solid preform of prescribed compositional and layer structure of core and cladding; and securing of the fibre which is pulled from the preform by the use of an electric furnace.

It is to be noted that the preform is 7 mm or more in diameter, while the fibre diameter is $125 \pm 1\,\mu m$. As we are concerned with the structural characteristics of optical fibres, it would be informative to give an account of the MCVD method for production of the preform with its layer structure, which leads to the fibre structure. Figure 1.7 shows schematically the MCVD process. A fused quartz tube is mounted on a glass-working lathe and slowly rotated while reactants, usually $SiCl_4$ and dopant reactants $GeCl_4$ and BCl_4, flow through it in an oxygen stream. An oxyhydrogen burner is slowly traversed along the outside tube to provide simultaneous deposition and fusion of a layer of the reacting materials. Some fifty layers are deposited by multiple passes of the burner. To fabricate step-index fibres, the dopant concentration is held fixed as a function of the layer deposited, whereas for graded-index fibres it is gradually increased with increasing layer number. At the end of the process, the temperature of the burner is raised to collapse the tube into a solid preform.

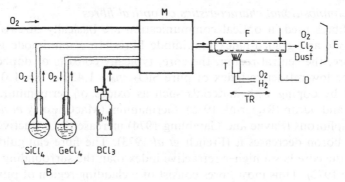

Figure 1.7 Schematic diagram of modified chemical vapour deposition process (MCVD). F, fused quartz tube; D, deposited layer of core glass; M, flow meters; B, bubblers; T, multiburner torch; TR, translation.

The last stage is to secure the fibre which is pulled from the preform. To examine the layer structure, Marcuse and Presby (1980) used scanning electron microscopy on an etched fibre end. Their results indicated the preservation of the layer structure by the fibre. Its scale in the fibre is of the order of less than one wavelength. In addition, the refractive index is not uniform within each layer.

The structural characteristics of preforms and optical fibres fabricated by MCVD were also studied by Presby *et al* (1975), applying optical inter-ference, the slab method described in Chapter 3 and scanning electron microscopy. They reported that structural features resulting from the deposition process are preserved through the subsequent processing and appear in the fibre, in addition to the occurrence of a central dip. The preservation of the deposited profile is based on the observation of a linear increase in refractive index in the fibre which was pulled from the preform in which the dopant concentration was increased in the same manner.

A slice transverse to the axis of the preform was polished to a thickness of about 10 mm for interference and optical microscopy. Transverse samples of the fibre were also prepared for interference microscopy and scanning electron microscopic investigations (Burrus and Standley 1974). Structural characteristics of the preform were resolved by both optical and inter-ference microscopy.

A two-beam interferogram of a preform slab (Presby *et al* 1975) showed straight parallel fringes in the fused silica cladding, indicating its uniform composition, followed by a drop in the level of the fringes, indicating the start of the borosilicate step that has a lower index of refraction than pure fused silica. This layer prevents impurity diffusion into the core. Core deposition results from increasing flow of $GeCl_4$, thus producing an increasing GeO_2 content and associated increased refractive index with increasing dopant thickness. Presby *et al* increased the $GeCl_4$ flow eleven times, in such a manner as to produce a near-parabolic index variation from the cladding interface to the core centre. Their interferogram revealed the layer structure within the core region corresponding to each torch traversal. Quantitative index measurements indicated a maximum index difference between core and cladding of $\Delta n = 0.016$, which is in good agreement with measurements from interferograms on fibres. They also reported that resolution of the interference microscope was not sufficient to resolve any layer or step structure in the fibre and they successfully used scanning electron beam microscopy to resolve layer structure in fibres.

Barakat *et al* (1988) applied multiple-beam Fizeau fringes to graded-index optical waveguides to examine fibre formation. The multiple-beam inter-ference system revealed the existence of successive layers from the graded-index fibre core. Both thickness and refractive index, graded from one layer to another, have been estimated. Figure 1.8 shows an interferogram resolving the layer structure. The fibre is found to be formed of a succession of step-index layers; $n(r)$ remains constant over the thickness Δr, yet it follows the known function relating $n(r)$ with r in terms of $n_{r=0}$ and α, namely

$$n(r) = n_0 \left[1 - 2 \, \Delta \left(\frac{r}{a} \right)^{\alpha} \right]^{1/2} \qquad 0 \leqslant r \leqslant a \qquad (1.1)$$

where r is the distance from the core centre, a the core radius, $\Delta = (n^2(0) - n^2(a))/2n^2(0)$ and α is a parameter characterising the shape of the index profile. The index profile for graded-index fibres showing such structure is presented theoretically and verified experimentally from micro-interferograms. The radius of the core is subdivided into m layers or zones of width Δr. r_m designates the order radius of the mth layer

$$0 = r_0 < r_1 < r_2 \ldots < r_{m-1} < r_m = a$$

$$(r_m - r_{m-1}) = \frac{a}{m} = \Delta r = \text{constant}.$$

The refractive index $n(0) = n_{r_0} > n_{r_1} > \ldots > n_{r_m} = n_a = n_{\text{clad}}$ and $n(r) = f(r)$, the basic equation of the graded-index core. For all m, r_m is obtained by substituting in equation (1.1). The magnitude of the fringe shift resulting from the contribution of the m layers constituting the core, in addition to the cladding, is obtained by summation. Each layer contributes half an ellipse of semi-principal axes:

$$\left\{ r_f, \frac{4 \, \Delta Z}{\lambda} (n_{\text{clad}} - n_L) r_f \right\}, \left\{ a, \frac{4 \, \Delta Z}{\lambda} (n_{r_{m-1}} - n_{\text{clad}}) a \right\},$$

$$\left\{ r_{m-1}, \frac{4 \, \Delta Z}{\lambda} (n_{r_{m-2}} - n_{r_{m-1}}) r_{m-1} \right\}, \ldots$$

as explained in Chapter 4.

Figure 1.8(a) shows a microinterferogram of a graded-index fibre, $t_f = 125 \pm 1$ μm, core radius $a = 25$ μm, $n_L = 1.4623$ and $n_{\text{clad}} = 1.4597$ for $\lambda = 5461$ Å. The multiple-beam fringes in the liquid region are straight lines parallel to the edge of the silvered liquid wedge with the fibre immersed, with interfringe spacing ΔZ. As they cross the liquid/clad interface, they follow part of an ellipse, except for the case of a matching liquid when $n_L = n_{\text{clad}}$. Both fringe systems are continuous with a discontinuity at the liquid/clad interface. As the fringes cross the core region, they show successive discontinuities. These result from abrupt changes in refractive index profile due to the presence of a sequence of deposition layers for a particular dopant concentration. It is to be noted that successive dis-continuities in fringes exist only in the core region and for all orders of interference which appear at different points along the fibre length and across the core. The same effect has been observed using liquids of different n_L (figure 1.8(b)). The fringe points on the (Z, X) plane where the fringe system is formed, allow the calculation of both α and Δn by the invariance method (Barakat *et al* 1985). The multistep index profile is as shown in

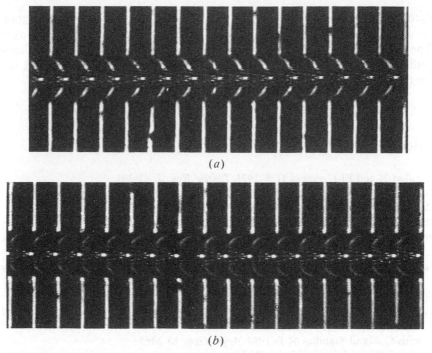

Figure 1.8 Multiple-beam microinterferogram resolving layer structure in the fibre core of a graded-index fibre, $t_f = 125\ \mu m$, core radius $= 25\ \mu m$ and $n_{clad} = 1.4597$ for the wavelength $\lambda = 5461$ Å. $n_L = 1.4623$ for the case (a) and $n_L = 1.4588$ for the case (b).

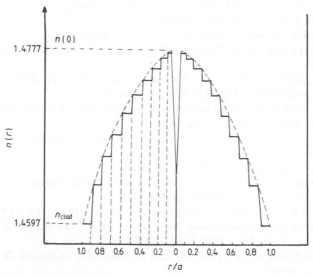

Figure 1.9 Diagrammatic presentation of step-pyramid-like index profile with a central dip. $\Delta n = 0.018$.

figure 1.9. The extension of each layer is determined from the interferogram on the (Z, X) plane by the projection of the fringe segment on the X axis. The layer extension $r_m - r_{m-1} = \Delta r$ is found to be $0.1a = 2.5$ μm, the core radius $a = 25$ μm and $m = 10$; $(n_{r_{m-1}} - n_{r_m})$ decreases in steps towards the centre of 0.0020.

References

Barakat N and El-Hennawi H A 1971 *Textile Res. J.* **41** 391
Barakat N, El-Hennawi H A and El-Diasti F 1988 *Appl. Opt.* **27** 5090
Barakat N, Hamza A A and Goneid A S 1985 *Appl. Opt.* **24** 4383
Barakat N and Hindeleh A M 1964 *Textile Res. J.* **34** 581
Bunn C W 1949 The optical properties of fibres in *Fibre Science* ed. J M Preston (Manchester: The Textile Institute) pp144–57
—— 1961 *Chemical Crystallography—An Introduction to Optical and X-Ray Methods* (Oxford: Oxford University Press) pp304–22
Bunn C W and Daubeny P 1954 *Trans. Faraday Soc.* **50** 1173
Burrus C A, Chinnock E L, Gloge D, Holden W S, Tingue Li, Standley R D and Keck D B 1973 *Proc. IEEE* **61** 1498
Burrus C A and Standley R D 1974 *Appl. Opt.* **13** 2365
Carter G B and Schenk V T J 1975 Ultra-high modulus organic fibres in *Structure and Properties of Oriented Polymers* ed. I M Ward Ch. 13 (London: Applied Science) pp454–92
Costa B 1980 The optical fibre in *Optical Fibre Communication* Technical Staff of CSELT (New York: McGraw-Hill) pp1–46
Denbigh K G 1940 *Trans. Faraday Soc.* **36** 936
Dobb M G and McIntyre J E 1984 *Adv. Polym. Sci.* **60/61** 61
Faust R C 1952 *Proc. Phys. Soc.* **B65** 48
Françon M 1961 *Progress in Microscopy* (Oxford: Pergamon)
French W G, Pearson A D, Tasker G W and MacChesney J B 1973 *Appl. Phys. Lett.* **23** 338
Hamza A A and Kabeel M A 1986 *J. Phys. D: Appl. Phys.* **19** 1175
Hamza A A and Sikorski J 1978 *J. Microsc.* **113** 15
Hamza A A, Sokkar T Z N and Kabeel M A 1985a *J. Phys. D: Appl. Phys.* **18** 1773
—— 1985b *J. Phys. D: Appl. Phys.* **18** 2321
Hartshorne N H and Stuart A 1970 *Crystals and the Polarizing Microscope* (London: Edward Arnold) pp556–88
Keller A 1968 *Rep. Prog. Phys.* **31** 623
Kita H, Kitano I, Uchida T and Furukawa M 1971 *J. Am. Ceramic Soc.* **54** 321
Kuhn W and Grün F 1942 *Kolloidzchr.* **101** 248
MacChesney J B, O'Connor P B and Presby H M 1974 *Proc. IEEE* **62** 1280
Marcuse D 1972 *Light Transmission Optics* (New York: Van Nostrand Reinhold)
Marcuse D and Presby H M 1980 *Proc. IEEE* **68** 668
Meredith R and Hearle J W S 1959 *Physical Methods of Investigating Textiles* (New York: Interscience)

Miya T, Terunuma Y, Hosaka T and Miyashita T 1979 *Electron. Lett.* **15** 106

Northolt M G 1974 *Europ. Polym. J.* **10** 799

Payne D N and Gambling W A 1974 *Electron. Lett.* **10** 289

Presby H M, Standley R D, MacChesney J B and O'Connor P B 1975 *Bell Syst. Tech. J.* **54** 1681

Preston J M 1933 *Trans. Faraday Soc.* **29** 65

Rigterink M D 1975 *Tech. Dig. Topical Meet. Optical Fibre Transmission Jan. 7–9, Williamsburg, VA*

Stoves J L 1957 *Fibre Microscopy* (London: National Trade Press)

Wells O C 1974 *Scanning Electron Microscopy* (New York: McGraw-Hill)

2 Principles of Interferometry

2.1 Introduction

Consider the light disturbances at two points P_1 and P_2 in a wavefield produced by an extended quasi-monochromatic source. Let us assume that the wavefield is in a vacuum and that P_1 and P_2 are many wavelengths away from the source. When P_1 and P_2 are close enough to each other, the fluctuations of the amplitudes at these points and also the fluctuations of the phases will not be independent. It is reasonable to suppose that if P_1 and P_2 are so close that the difference in path (PD) $= SP_1 - SP_2$ from each point S is small compared to the mean wavelength $\bar{\lambda}$, then the fluctuations at P_1 and P_2 will effectively be the same. Also some correlation between the fluctuations will exist even for greater separation of P_1 and P_2 provided that for all source points the PD does not exceed the coherence length $C \ \Delta t = C/\Delta \nu = (\bar{\lambda})^2/\Delta \lambda$. We are thus led to the concept of a region of coherence around any point P in a wavefield.

In order to describe adequately a wavefield produced by a finite poly-chromatic source, it is evidently desirable to introduce some measure for the correlation that exists between the vibrations at different points P_1 and P_2 in the field. We must anticipate that such a measure would be closely linked to the sharpness of the interference fringes which would result on combining the vibrations from the two points. We should expect sharp fringes when the correlation is high, e.g. when the light at P_1 and P_2 comes from a very small source of a narrow spectral range $\Delta \nu$. We should also expect to have no fringes in the absence of correlation, e.g. when P_1 and P_2 each receives light from a different physical source.

2.2   Division of the wavefront

There are many ways of dividing the wavefront into two segments and recombining them at a small angle. Interference fringes formed by two slits in Young's experiment, the Fresnel mirror and the Fresnel biprism are examples of division of the wavefront, where two beams derived from the same source follow two separate paths with different optical lengths and are allowed to meet as shown in figure 2.1.

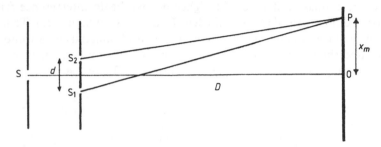

Figure 2.1   Young's double-slit experiment. S is a monochromatic light source.

The phase difference δ between the two intersecting beams is equal to $2\pi/\lambda \times$ path difference. When $\delta = 2m\pi$, the path difference equals $m\lambda$, m being $0, 1, 2, 3, \ldots$, and bright fringes are formed and located on a screen at $x_m = m\lambda D/d$ where d is the separation of the two slits and D the distance of the screen from the plane of the double slits S_1 and S_2. In all these examples, the intensity distribution in the fringe pattern follows a cosine square law, being equidistant and non-localised, i.e. formed in space on any plane in front of the two slits:

$$I = 4I_0 \cos^2 \delta/2.$$

I_0 is the intensity of the two separate waves, as shown in figure 2.2.

Dark fringes are formed and located at $x = (m + \frac{1}{2})\lambda D/d$, Δx being the separation between any two successive dark or bright fringes.

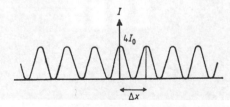

Figure 2.2   Intensity distribution in the case of two-beam interference.

The visibility of fringes decreases as the path difference increases since the source is never strictly monochromatic. The visibility vanishes at high orders of interference, away from the point O (figure 2.1), when the path difference exceeds the coherent length of the source of illumination.

Another example of interference fringes resulting from division of the wavefront is to be found in Rayleigh's refractometer where monochromatic light from a linear source is made parallel by a lens L_1 and split into two beams by two slits S_1 and S_2 at a distance apart. The two beams follow two exactly equal metric lengths, but one of them passes through a phasor which introduces a phase difference. Straight-line two-beam interference fringes are formed on the focal plane of the lens L_2 as shown in figure 2.3. To locate the centre of the system, the zero-order fringe is obtained using a white light source and the two systems are seen and recorded simultaneously, each of them covering half the field of view as shown in figure 2.4.

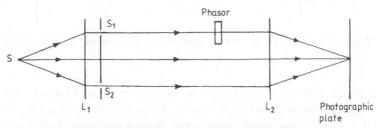

Figure 2.3   Rayleigh's refractometer.

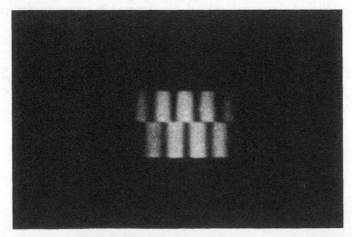

Figure 2.4   The zero-order fringe (for explanation of the production of these fringes see text).

2.3 Division of amplitude

Of equal importance when considering practical applications are the interference effects resulting from division of amplitude. When light falls on a thin parallel-sided film, in air, of refractive index n, part of the amplitude of the incident wave is reflected at the first interface, air/film, and the remainder traverses the film after refraction to meet the second interface as shown in figure 2.5. At the point B, reflection takes place at the film/air interface and part is transmitted along BE. A similar fraction traverses the film to the first interface where reflection film/air takes place and a fraction is transmitted at the point C. The two fractions leaving the first interface at A and C are parallel and when brought to intersect at the focal plane of a lens produce interference fringes at reflection, since they are coherent. Also, the two rays leaving the second interface of the film at B and D, when brought to intersect at the focal plane of a lens, produce two-beam interference in transmission, as they are also coherent. Let us calculate the path difference between the two rays brought to intersect on the focal plane of the lens forming fringes in transmission. As the refracted ray reaches the point B, it is divided into two parts, one is transmitted along BE and the other traverses the route BCD. As the disturbance reaches the point D, the other disturbance along BE would have travelled to the point D′ where BD′ = nBC + nCD. These two disturbances were originally in phase as originated from the same point B. A lens brings all disturbances with their phases on a plane perpendicular to its axis to meet at its focal plane. Accordingly we are concerned with the phase difference between the two disturbances at D and E. Since the two disturbances at D and D′ are in phase, the phase difference between the disturbances at D and E is equal to

$$\frac{2\pi}{\lambda}\,\mathrm{ED}' = \frac{2\pi}{\lambda}\,(n\mathrm{BC} + n\mathrm{CD} - \mathrm{BE}) = \frac{2\pi}{\lambda}\,2nt\cos r$$

with notation as in figure 2.5. For bright fringes in transmission

$$\frac{2\pi}{\lambda}\,2nt\cos r = 2m\pi$$

$$m\lambda = 2nt\cos r$$

m being the order of interference. At reflection, since the change of phase on reflection at the air/film interface equals π, the reflected ray at A suffers a change of phase of π with respect to that transmitted and reflected at B, film/air. The condition for a bright fringe at reflection is

$$(m + \tfrac{1}{2})\lambda = 2nt\cos r.$$

The change of phase on reflection at the film/air interface is zero. On illuminating the film with monochromatic light, interference fringes of successive orders of interference are formed when film thickness t is kept constant but with a variable angle of incidence θ. The fringes are fringes of equal θ, i.e. of equal inclination. They are localised at infinity and brought in focus on the focal plane of a lens as shown in figure 2.5. When θ is kept constant t must vary to obtain fringes of successive orders of interference. These fringes are termed the fringes of equal thickness. They are localised at a definite plane in space, and in the case of an air wedge illuminated by a parallel beam of monochromatic light making an angle of incidence θ, the fringes are straight lines parallel to the edge of the wedge. Their plane of localisation is very close to the wedge for normal incidence.

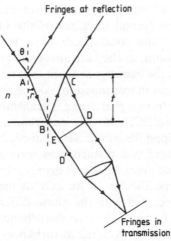

Figure 2.5   Interference fringes formed in transmission and at reflection by division of amplitude.

In general the fringes are of equal nt if n is position dependent. They are the fringes of equal optical thickness.

Another example of interference fringes resulting from division of amplitude is to be found in the Michelson interferometer. Here the two beams obtained by amplitude division follow two different directions against plane mirrors, whence they are brought together again to form interference fringes.

2.4   Interference of plane polarised light

Fresnel and Arago (see Tolansky 1948) discovered experimentally the rules required for two polarised beams to produce interference fringes. These rules are:

(*a*) two rays of light polarised at right angles to each other cannot be made to interfere;

(*b*) two rays of light plane polarised in parallel planes can be made to interfere;

(*c*) two plane polarised beams brought into the same plane will interfere if they are emerging from the same source, i.e. coherent.

We shall confine ourselves to a few simple cases of relevance. The first is the case of a parallel plate of (i) uniaxial material and (ii) biaxial material. The second case is of some natural and synthetic fibres that show double refraction, i.e. birefringence.

It follows from rule (*b*) that multiple rays of plane polarised light vibrating in the same plane can be made to interfere, producing multiple-beam interference fringes.

2.4.1 The case of a uniaxial crystal cut perpendicular to the optic axis

Unless travelling along the optic axis, an incident beam entering the crystal is split into two beams polarised in perpendicular planes.

(*a*) The ordinary ray whose electric vector is vibrating at right angles to the plane of incidence, with a constant refractive index n_o, which is independent of the direction of propagation.

(*b*) An extraordinary ray whose electric vector is vibrating in the plane of incidence, with a refractive index n_e' that varies with the angle of incidence. This refractive index attains a limiting value n_e for light incident perpendicular to the optic axis, which is given by the following equation for any angle of refraction r_e:

$$\frac{1}{n_e'^2} = \frac{\cos^2 r_e}{n_o^2} + \frac{\sin^2 r_e}{n_e^2}.$$

Multiple reflected rays resulting from each of the two refracted rays produce a system of interference fringes. If δ_1 and δ_2 are the path differences for the two systems

$$\delta_1 = 2n_o t \cos r_o$$

and

$$\delta_2 = 2n_e' t \cos r_e$$

neglecting the deviation from the basic formula.

At the centre of the two systems, i.e. at $\theta = 0$, $n_e' = n_o$. Consequently no separation takes place between the members of the same order of interference belonging to each system. As θ increases, n_e' differs steadily from n_o and separation takes place which increases with θ.

These interference systems can easily be formed using a thin sheet of freshly cleaved high-quality phlogopite mica. The mica is doubly silvered by coating each face with a thin layer of silver of reflectivity $R \approx 80\%$ by evaporation. Fringes of equal chromatic order (see Tolansky 1948) are obtained, and are shown in figure 2.6. No birefringence doubling effect is detected using a constant-deviation spectrograph. This establishes that the sample of phlogopite mica can be considered a uniaxial crystal.

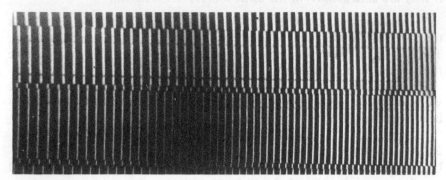

Figure 2.6   Fringes of equal chromatic order for a uniaxial crystal.

The specimen was then bent in the form of half a cylinder and a parallel beam of monochromatic light was allowed to be incident. Straight-line multiple-beam interference fringes parallel to the axis of the cylinder are formed in focus on a photographic plate placed normal to the cylinder axis and passing by the centre O. Variation of the angle of incidence occurs from $0°$ at the centre to close to grazing. They are the fringes of equal tangential inclination (Tolansky and Barakat 1950). The resulting fringes appear in figure 2.7.

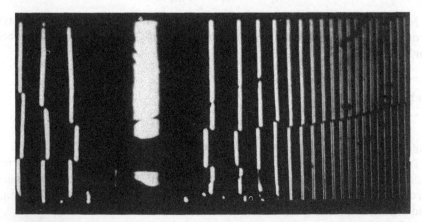

Figure 2.7   Fringes of equal tangential inclination for a uniaxial crystal.

The introduction of a Nicol prism into the incident beam proved that the two systems of fringes are vibrating at right angles to each other, the outer fringes being formed by the beam whose electric vector vibrates perpendicular to the plane of incidence, i.e. the ordinary ray. The reason lies in the fact that phlogopite mica is a negative crystal, i.e. $n_o > n_e$ (Barakat 1958).

2.4.2 The case of a biaxial crystal cut perpendicular to the acute bisectrix

Since any ray incident on the crystal is generally split into two rays polarised in perpendicular planes and travelling with slightly different velocities and in different directions, the multiply reflected rays resulting from each of the refracted rays will produce a system of interference fringes. If δ_1 and δ_2 are the path differences for the two systems we have

$$\delta_1 = 2n^\perp t \cos r'$$

and

$$\delta_2 = 2n^\| t \cos r''$$

where $\sin \theta = n^\perp \sin r' = n^\| \sin r''$, $n^\|$ and $n^\perp$ being the refractive indices for the two refracted rays for a certain angle of incidence θ.

Let us consider the plane containing the optic axis. The section of the wave surface by that plane consists of a circle of radius n_m and an ellipse where axes are n_g and n_p. For the circular section

$$\delta_1 = 2n_m t \cos r$$

and

$$\delta_1^2 = 4t^2(n_m^2 - \sin^2 \theta). \tag{2.1}$$

For the elliptic section

$$\frac{1}{n^2} = \frac{\cos^2 r}{n_g^2} + \frac{\sin^2 r}{n_p^2}$$

therefore

$$n^2 = n_g^2 - \sin^2 \theta \left(\frac{n_g^2}{n_p^2} - 1 \right) \tag{2.2}$$

and

$$\delta_2^2 = 4t^2 \left(n_g^2 - \frac{n_g^2}{n_p^2} \sin^2 \theta \right). \tag{2.3}$$

Accordingly, two independent systems of fringes polarised in mutually perpendicular planes are formed. The first system corresponds to equation (2.1), is of constant refractive index n_m and vibrates perpendicular to the plane of incidence which is also the optical axial plane. The other corresponds to equations (2.2) and (2.3) with variable refractive index n, with angle θ as given by equation (2.2). The refractive index n is equal to n_g at $\theta = 0$ where $n_g > n_m > n_p$. As θ increases n decreases and attains a value n_m at an incidence equal to the apparent optic axial angle E satisfying the equation

$$\sin E = n_p \left(\frac{n_g^2 - n_m^2}{n_g^2 - n_p^2}\right)^{1/2}.$$

It decreases further to the limiting value n_p at $\theta = \pi/2$.

This means that, starting from small angles of incidence, the outer fringe is that vibrating in the plane of incidence and the separation between each pair of fringes belonging to the two systems and of the same order of interference decreases with θ until they overlap at $\theta = E$. As the angle of incidence increases beyond $\theta = E$ this separation increases, but the inner fringes now vibrate in the plane of incidence.

Experimentally, a piece of high-quality, muscovite mica was selected and the plane containing the optic axes was determined by means of a polarising microscope. A thin film was cleaved and doubly silvered. It was then bent in the form of a cylinder along the direction of the intersection of the optical axial plane with the cleaved surface. The resulting fringes of equal tangential inclination are shown in figure 2.8. The two systems are polarised in perpendicular planes. Before overlapping occurs the outer fringe corresponds to the ray vibrating in the plane of incidence. After overlapping, the inner system corresponds to this vibration direction (Barakat 1958).

Figure 2.8   Fringes of equal tangential inclination for a biaxial crystal.

2.4.3   *The case of a synthetic fibre exhibiting birefringence*
When Acrilan fibre is introduced in a silvered liquid wedge and a parallel beam of monochromatic light illuminates the interferometer, the incidence being normal, multiple-beam Fizeau fringes are formed in transmission and

at reflection. Figure 2.9 shows multiple-beam Fizeau fringes in transmission for plane polarised light vibrating parallel and perpendicular to the fibre axis. A resolved doublet results from birefringence of the Acrilan fibre (Barakat and El-Hennawi 1971). Here we are only quoting Acrilan fibre as an example to illustrate interference of plane polarised light, but a detailed account of the application of multiple-beam fringes to the determination of refractive indices and birefringence of fibres is given in Chapter 4.

Figure 2.9 A microinterferogram of multiple-beam Fizeau fringes in transmission crossing an Acrilan fibre. (From Barakat and El-Hennawi 1971.)

2.5 Holography and holographic interferometry

2.5.1 Holography

The foundation of holography was laid by Gabor (1948). Holography records complete information about a wave—both in amplitude and phase. Apart from amplitude information, which is recorded as intensity by a photographic plate, phase information is recorded by the superposition of a coherent reference wave. In practice, holograms of objects were achieved in 1962 after the invention of lasers. Accordingly a hologram is an interference pattern formed by an object wave and a reference wave. Figure 2.10 shows schematically a set-up for preparing a hologram. Light waves of complex amplitude from an object $A_o = a_o \exp(iQ_o)$ and from a reference $A_r = a_r \exp(iQ_r)$ fall on a photographic plate. These waves are coherent, they form an interference pattern on the photographic plate in which the distribution of intensity $I_{(x,y)}$ is given by

$$I_{(x,y)} = |A_o + A_r|^2 = (A_o + A_r)(A_o^* + A_r^*)$$
$$= a_o^2 + a_r^2 + A_o A_r^* + A_o^* A_r.$$

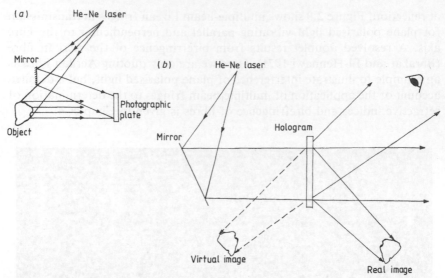

Figure 2.10 Schematic diagram for recording a hologram (*a*) and reconstruction of the wavefront (*b*).

2.5.2 Holographic interferometry

The applications of holography are classified into two major domains: applications requiring three-dimensional images for visual perception, and applications in which holography is used as a measuring tool.

When viewing a three-dimensional holographic picture, the hologram will register the most minute details, such as cracks, ledges and surface roughness of about one micrometre in size (see Denisyuk 1978).

One of the important techniques in which holography is employed as a kind of measuring tool is holographic interferometry (Ostrovsky *et al* 1980, Vest 1979). The general idea of the method is illustrated in figure 2.11. The same holographic plate H is used to record holograms of the object in its initial state O_1 and in its final state O_2, for example an object deformed under a load *P*. In this double-exposure technique a hologram of the undisturbed object is made, then, before processing, one exposes the hologram for a second time to the light coming from the new distorted object. The result is two overlapping reconstructed waves which proceed to form a fringe pattern indicative of the displacements suffered by the object, i.e. change in optical path length. Variations in index such as those arising in wind tunnels and similar mechanisms will generate similar patterns. The accuracy of measurements of distance changes of a tenth of a micrometre or less can be distinguished.

One of the applications of holographic interferometry is to be found in aerodynamics where it is used to investigate the flow around various bodies.

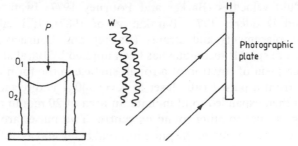

Figure 2.11 Recording holograms by the double-exposure technique.

2.6 Speckle and speckle interferometry

2.6.1 Speckle

When a rough surface is illuminated by coherent light, the reflected beam consists of random patterns of bright and dark regions known as speckle (Françon 1979). These patterns can be interpreted in terms of Huygen's principle; the intensity at any field point is caused by the interference of wavelets, scattered from different points within the illuminated area, with their phases randomised by height variations on the rough surface. The spatial pattern and contrast of the speckle depend on the optical system used for observation, the condition of coherence of the illuminating beam and the surface roughness of the scatterer.

Speckle may be observed when a coherent beam of monochromatic light is allowed to transmit through a partially transparent plate with a rough surface. Figures 2.12(*a*) and (*b*) show a simple optical arrangement for recording speckle in transmission and at reflection. The speckle pattern is found in the whole of the space in front of the optically rough surface and the screen AB. For a circular illuminated area of diameter D illuminated by coherent light of wavelength λ, the average size $\langle \delta \rangle$ of the speckle is given by the appropriate formula

$$\langle \delta \rangle = 1.2L/D$$

L being the distance between the illuminated rough surface and the screen or photographic plate. The number of scatterers which contribute to the interference producing a speckle pattern is an important factor that determines the properties of the pattern.

In speckle pattern photography a diffuse object illuminated by a coherent beam is imaged twice. The speckle pattern formed in one image is displaced by a few micrometres. The pattern is always correlated to that of the second image. This is termed double-exposure speckle photography. Speckle pattern photography and interferometry techniques have been widely used in measuring fluid velocities (Barker and Fourney 1977, Iwata *et al* 1978, Simpkins and Dudderer 1978, Barakat *et al* 1987). It is also used in measuring displacements and strains (Barker and Fourney 1976). The double-exposure speckle technique has been applied by Barakat *et al* (1986) to the measurement of rotation of a rough surface in the form of a ground glass. Light from a pulsed ruby laser at $\lambda = 6940$ Å is directed through a pin-hole and then expanded to illuminate an area of 20 mm in diameter on the ground glass disc rotating round its centre. Two pulses are taken on a fine-grain photographic plate, the pulse interval being 800 μs. After processing the photographic plate is placed in the filtering system and a system of Young's fringes is observed. The fringe spacing s is related to the displacement by $x = D/s$, D being the distance between the photographic plate and the plane of observation.

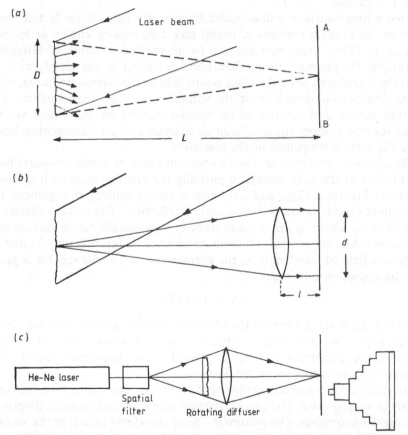

Figure 2.12 The optical arrangement for recording speckle. See text for details.

2.6.2 Speckle interferometry

In speckle interferometry (Jones and Wykes 1983), a system of interference fringes that results from fluctuations in the correlation between the two speckle patterns is formed. This is achieved by either a translation between the correlated portions of the patterns or the existence of a phase variation between them. The optical set-up used in speckle interferometry is as shown in figure 2.12(c). The light from an He–Ne laser is allowed to be incident on a rough ground glass surface through two circular apertures of 2 mm in diameter, 20 mm apart. A lens is used to form the speckle image of the object with unit magnification. The spacing of grid structure resulting within speckles is found to be equal to 9.5 μm. Two exposures of the formed speckle with its grid structure are recorded with a displacement of the object equal to double the grid spacing. This is followed by recording the speckles after object displacement of multiples of the grid spacing namely 38, 57 and 76 μm. The displacement direction is parallel to the line joining the two apertures. After developing, the doubly exposed plate is placed in the filtering set-up shown in figure 2.13. The grid structure formed is shown in figure 2.14 (Barakat *et al* 1988). For the measurement of displacement and consequently velocities, it is informative to state the range of displacements that could be measured by the double-exposure technique applying speckle pattern photography and applying speckle interferometry. When using speckle pattern photography, the least measurable displacement is equal to the minimum speckle size δ formed by a certain rough object used as a diffuser. From the specification of the optical set-up used the order of the speckle size is about ten micrometres. To record and resolve such speckle experimentally a high-resolution emulsion has to be used; the grain size of the emulsion should be less than the speckle size.

For object displacement smaller than the speckle size, double-aperture speckle interferometry should be used. The smallest measurable displacement is equal to the grid spacing formed within the speckle. This spacing is equal to $\lambda p/D$ where D is the separation between the two identical apertures and p is the object distance.

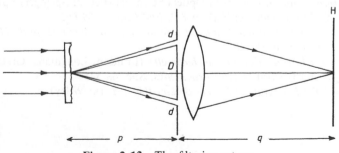

Figure 2.13 The filtering set-up.

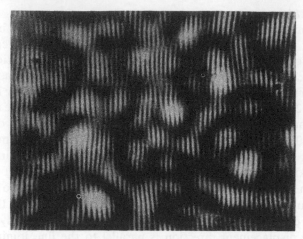

Figure 2.14   Grid structure formed by laser speckle interferometry.

References

Barakat N 1958 *J. Opt. Soc. Am.* **48** 92
Barakat N, El-Ghandoor H, Merzkirch W and Wernekink U 1988 *Exp. Fluids J.* **6** 71
Barakat N and El-Hennawi H A 1971 *Textile Res. J.* **41** 391
Barakat N, Hamed A H and El-Ghandoor H 1987 *Optik* **76** 78
Barakat N, Merzkirch W and El-Ghandoor H 1986 *Optik* **74** 114
Barker D B and Fourney M E 1976 *Exp. Mech.* **18** 209
—— 1977 *Opt. Lett.* **1** 136
Denisyuk Yu N 1978 *Fundamentals of Holography* (Moscow: Mir) pp116–18
Françon M 1979 *Laser Speckle and Applications in Optics* (New York: Academic)
Gabor D 1948 *Nature* **161** 777
Iwata K, Hakoshima T and Nagata R 1978 *Opt. Commun.* **25**
Jones R and Wykes C 1983 *Holographic and Speckle Interferometry* (Cambridge: Cambridge University Press)
Ostrovsky Yu I, Butusov M M and Ostrovskaya G V 1980 *Interferometry by Holography* (Springer Series in Optical Sciences) (Berlin: Springer) pp73–5
Simpkins P G and Dudderer T D 1978 *J. Fluid Mech.* **89** 665
Tolansky S 1948 *Multiple-Beam Interferometry of Surfaces and Thin Films* (Oxford: Clarendon) p126
—— 1955 *An Introduction to Interferometry* (London: Longmans, Green)
Tolansky S and Barakat N 1950 *Proc. Phys. Soc.* **63** 545
Vest C M 1979 *Holographic Interferometry* (New York: Wiley)

3 Two-beam Interferometry Applied to Fibrous Materials

3.1 Introduction

When discussing fibre structure in Chapter 1, it became obvious that optical anisotropy exists in most natural and synthetic fibres. In these fibres, the refractive indices for plane polarised light vibrating parallel and perpendicular to the fibre axis, $n^{\parallel}$ and $n^{\perp}$, and the birefringence (double refraction) $\Delta n = n^{\parallel} - n^{\perp}$, are the parameters that determine optical anisotropy and characterise the structure of the fibre on the molecular level. The determination of such parameters provides useful information as they play an important role in the elucidation of the molecular arrangement within the fibre. These optical properties can be determined using two-beam interference microscopes. They provide quantitative information about optical properties of the skin and core of heterogeneous fibres. Information on the variation of refractive indices of the fibre with (i) wavelength of light, (ii) temperature and (iii) tensile strength is available using two-beam interference microscopy.

The aim of this chapter is to present the theory and application of two-beam interference microscopes to the study of fibrous materials.

(*a*) Fibres with regular and irregular transverse sections
(i) homogeneous fibres;
(ii) heterogeneous fibres with skin/core structure;
(iii) multilayer fibres.
(*b*) Optical fibres, namely step-index and graded-index fibres.

Interferometric determination of the surface topography of fibres, degree of smoothness and irregularities in the fibre's diameter along the fibre extension is dealt with in Chapter 5.

Objects in optical microscopy can be classified as amplitude or phase objects. Amplitude objects vary in their light absorption with respect to the medium that surrounds them, thus exhibiting a certain amount of contrast. Phase objects produce no variations in light being absorbed and differ from the surrounding medium merely by their optical thickness nt, n being the refractive index and t the object's metric thickness. In interference microscopy we are primarily concerned with phase objects. Many optical microscopes were developed using the two-beam interference technique; the Mach–Zehnder interferometer, the Nomarski interferometer, the Pluta polarising interference microscope, the Interphako interference microscope and the Baker, Dyson, Leitz and Zeiss–Linnik interference microscopes. A detailed account of the optical arrangement, passage of light and experimental procedure for the formation of two-beam interferograms of phase objects using these interference microscopes is given in Chapter 7. In the present chapter we discuss the theory and application of these microscopes to the determination of fibre characteristics, with emphasis on refractive indices, of regular and irregular fibres of homogeneous and heterogeneous transverse sections. Also, index profiles of step-index and graded-index optical fibres are elucidated from two-beam interference fringe patterns.

3.1.1   Previous investigations and reviews of the literature on the development of interference microscopy and its application to textile fibre materials

Among investigators, Pluta (1982) (see the book review by Sikorski (1984)), gave a review of the development of interference microscopes, beginning with the design of Jamin–Lebedev, Smith, Pluta and the microscopes following the Mach–Zehnder system. Two-beam interference microscopes, including microscopes with reflected light, are discussed in detail. The reasons underlying the design of the differential interference contrast microscope (Pluta 1971, 1972) are given. The two-beam interference microscope (Smith–Baker type) was used by Faust (1956) to determine refractive index variations within optically heterogeneous specimens. A set of reference fringes was produced in the background by inserting a quartz plate between the fibre and the analyser. A modification of this technique was introduced by McKee and Woods (1967) in which the quartz plate was dispensed with. A Leitz two-beam interference microscope was used by McLean (1971) to analyse the anomalous differences in birefringence in polyester fibres. The birefringence of steam-stretched acrylic fibres was measured at different draw ratios, using the Zeiss ultraphot interference microscope (Blakey *et al* 1970). The Pluta interference microscope (Pluta 1965, 1971, 1972), which is based on double-refracting interference with a variable amount and direction of wavefront shear, is capable of giving either the uniform or fringe interference fields with a continuously variable amount and direction of lateral image duplication. Using this microscope, quantitative information about the refractive index and birefringence of the skin and core of the fibre can be secured. Pluta (1972) applied his double-refracting interference microscope to the study of synthetic polymer fibres. Hamza and Sikorski (1978) used the Pluta two-beam polarising interference microscope to determine the anisotropy of Kevlar (poly(p-phenylene terephthalamide)) fibres. They also calculated the principal polarisabilities of these fibres using a molecular model suggested by Northolt (1974), together with the values of bond polarisabilities given by Denbigh (1940) and also by Bunn and Daubeny (1954). Using the Lorentz −Lorenz relation they calculated the principal refractive indices of these highly oriented fibres. Simmens (1958) determined the birefringence in objects of irregular cross sectional shape. Hamza (1980) extended the applications of the Pluta microscope to measure the mean refractive indices and birefringence of fibres with irregular transverse sections. Żurek and Zakrzewski (1983) determined the refractive indices and birefringence of cotton fibres using the Pluta microscope, following the method described by Hamza (1980). The fibre is placed in an immersion liquid of high dispersion of refractive index and the measurements are carried out for two wavelengths of light. Two images of the fibre are observed, one with parallel refractive index and the other with perpendicular refractive index. The method described is in fact a combination of the Simmens method (1958) and the dual-wavelength method of Pluta (1965). Dorau and Pluta (1981a,b) dealt with the problem of correct measurement of optical path length (OPL) in the field of interference fringes, in terms of fringe displacements. Using the Pluta microscope with crossed polaroids white light can be used to identify the zero-order interference fringe, the achromatic black fringe. Hamza (1986), Hamza and Abd El-Kader (1986), Hamza and El-Farahaty (1986), Hamza and El-Dessouki (1987) and Hamza *et al* (1986) used the Pluta polarising microscope on fibres with regular and irregular transverse sections to determine the refractive indices and birefringence of polyester, Kevlar, cuprammonium and bicomponent fibres.

A comprehensive survey of investigators applying two-beam interference microscopy to textile fibres, the technique adopted and the type of synthetic or natural fibre investigated is given by Hamza (1986).

3.2 Theory of the Pluta microscope

This is a double-refracting interference microscope with variable amounts and direction of wavefront shear. The microscope is capable of giving either the uniform or fringe interference fields with continuously variable amounts and direction of lateral image duplication.

Hamza (1980) extended the application of the Pluta microscope to measure the mean refractive indices and birefringence of fibres with irregular transverse sections. Measurement of the mean refractive index of a fibre involves the thickness of both the skin and core regions. These measurements were carried out by the complementary use of scanning electron microscopy and interference microscopy. The scanning electron microscope was used to determine the cross sectional area (in mm^2) of the fibre.

If n_L is the refractive index of the immersion liquid and $n_a^{\parallel}$ and $n_a^{\perp}$ are the mean refractive indices of the fibre for plane polarised light vibrating parallel and perpendicular to the fibre axis respectively, then the optical path length differences, $\Delta\Gamma_{\parallel}$ and $\Delta\Gamma_{\perp}$, between the specimen and the immersion liquid are given by

$$\Delta\Gamma_{\parallel} = (n_a^{\parallel} - n_L)t \qquad (3.1)$$

$$\Delta\Gamma_{\perp} = (n_a^{\perp} - n_L)t \qquad (3.2)$$

where $\Delta\Gamma_{\parallel}$ and $\Delta\Gamma_{\perp}$ are in a length unit, e.g. mm, and t is the mechanical thickness of the object (fibre) in the same unit as $\Delta\Gamma$.

The birefringence, Δn_a, of the fibre is

$$\begin{aligned} \Delta n_a &= n_a^{\parallel} - n_a^{\perp} \\ &= (\Gamma_{\parallel} - \Gamma_{\perp})/t \end{aligned} \qquad (3.3)$$

Figure 3.1 gives the shape of fringes across a fibre of irregular cross sectional area A (or one of the duplicated images of the fibre using a shearing double-refracting interference microscope).

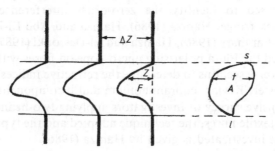

Figure 3.1   The shape of interference fringes across a fibre of irregular cross sectional area A.

The optical path length difference $\Delta\Gamma$ is given by the following formula:

$$\Delta\Gamma = \frac{Z}{\Delta Z}\lambda \qquad (3.4)$$

where Z is the interference fringe shift inside the fibre in a unit of length, e.g. mm, λ is the wavelength of monochromatic light used and ΔZ is the interfringe spacing. Therefore

$$\frac{Z^{\parallel}}{\Delta Z} = (n_a^{\parallel} - n_L) \frac{t}{\lambda} \tag{3.5}$$

$$\frac{Z^{\perp}}{\Delta Z} = (n_a^{\perp} - n_L) \frac{t}{\lambda} \tag{3.6}$$

$$Z^{\parallel} = t \frac{\Delta Z}{\lambda} (n_a^{\parallel} - n_L). \tag{3.7}$$

Integrating equation (3.7) in the region $s \geqslant d \geqslant l$ gives the area under the fringe shift $F^{\parallel}$:

$$\int_s^l Z^{\parallel} \, dx = \frac{\Delta Z}{\lambda} (n_a^{\parallel} - n_L) \int_s^l t \, dx$$

$$F^{\parallel} = \frac{\Delta Z}{\lambda} (n_a^{\parallel} - n_L) A \tag{3.8}$$

where A is the mean cross sectional area of the fibre. Therefore

$$n_a^{\parallel} = n_L + \frac{F^{\parallel}}{\Delta Z} \frac{\lambda}{A} \tag{3.9}$$

$$n_a^{\perp} = n_L + \frac{F^{\perp}}{\Delta Z} \frac{\lambda}{A} \tag{3.10}$$

$$\Delta n_a = \left(\frac{F^{\parallel} - F^{\perp}}{\Delta Z} \right) \frac{\lambda}{A}. \tag{3.11}$$

The direction of the fringe shift inside the fibre depends upon the relative values of the mean refractive index of the fibre and the refractive index of the immersion liquid.

As already mentioned, the Pluta interference microscope is capable of giving either uniform or fringe interference fields with a continuously variable amount and direction of lateral image duplication: two sheared (duplicated) images are obtained for one and the same object, with interference fringes deviated in such images in opposite directions (figure 3.2). It is more accurate to measure the double deviation of a fringe, $2Z$, instead of measuring the fringe deviation Z. Monochromatic light and white light are used with the Pluta microscope, the latter primarily to confirm the position of the achromatic fringe. Figure 3.3 shows duplicated images of a homogeneous fibre immersed in a liquid of refractive index n_L. When dealing with a fibre of circular cross section having a skin–core structure, the shapes of the interference fringes crossing the fibre are shown in figure 3.4. For the case of a homogeneous fibre with regular transverse section, the fringe shift produced for a circular fibre describes half an ellipse, having the two semi-major and minor axes: $a = r_f = t/2$ and $b = \delta Z$ at $x = 0$ (Barakat 1971). The area enclosed by the fringe shift between $x = -r_f$ to $x = +r_f$ is $F = \pi ab/2$ while the area A of the transverse section $= \pi t^2/4$, shown in figure 3.5.

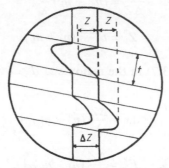

Figure 3.2 Totally duplicated images of the fibre; t is the fibre thickness, Z the fringe shift inside the fibre and ΔZ the interfringe spacing.

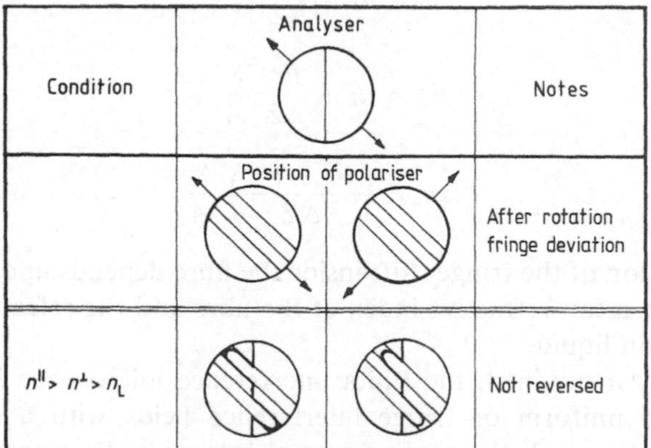

Condition	Analyser		Notes
	Position of polariser		After rotation fringe deviation
$n^{\parallel} > n^{\perp} > n_L$			Not reversed
$n^{\parallel} > n_L > n^{\perp}$			Reversed
$n_L > n^{\parallel} > n^{\perp}$			Not reversed

Figure 3.3 The direction of the fringe displacement across the fibre image with respect to the relative values of refractive indices $n^{\parallel}$ and $n^{\perp}$ of the homogeneous fibre and that of the immersion liquid n_L.

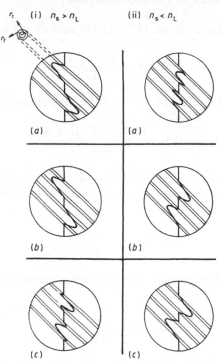

Figure 3.4 Duplicated images of the fibre, immersed in a liquid of refractive index n_L, using a double-beam polarising interference microscope. The shape of the interference fringe across a fibre of circular cross section with a core of radius r_c and refractive index n_c, and a skin of radius r_s and refractive index n_s, is shown. Two cases are considered: (i) when $n_s > n_L$ and (ii) when $n_s < n_L$. In each case there are three possibilities, (a) $n_c > n_s$ (b) $n_c = n_s$ and (c) $n_c < n_s$. (From Hamza 1986.)

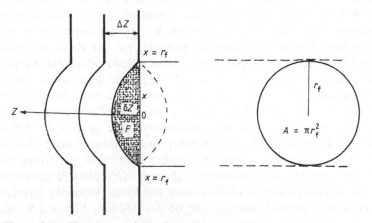

Figure 3.5 The shape of the interference fringes across a homogeneous cylindrical fibre of circular cross sectional area.

Substituting in the relations (3.9), (3.10) and (3.11) we have

$$n_a^\| - n_L = \frac{(\delta Z^\|)_{x=0}}{\Delta Z} \frac{\lambda}{t}.$$

A similar expression exists for $n_a^\perp$ and the birefringence

$$\Delta n_a = \frac{(\delta Z^\| - \delta Z^\perp)_{x=0}}{\Delta Z} \frac{\lambda}{t}.$$

This is the formula using a single-pass interferometer. For a double-pass interferometer, as in the case of a homogeneous fibre immersed in a silvered liquid wedge

$$\Delta n_a = \frac{(\delta Z^\| - \delta Z^\perp)_{x=0}}{\Delta Z} \frac{\lambda}{2t}$$

and for homogeneous fibres of irregular cross sections using a double-pass interferometer

$$n_a - n_L = \frac{F}{\Delta Z} \frac{\lambda}{2MA}$$

M being the magnification on the photographic plate (Sokkar and Shahin 1985, Wilkes 1985).

The accuracy of the measurement of the optical path length difference, with a fringe field (Wollaston prism), is about 0.05λ. Therefore the error in the determination of the refractive index and birefringence cannot be better than $0.003-0.001$ and the accuracy of the determination of the diameter of fibre is about 1 μm (Pluta 1972).

The following are illustrations given by microinterferograms to show the behaviour of the fringes as they cross the fibre and how information is extracted from these interferograms. Applications of two interference microscopes, the Pluta and the Interphako, to synthetic fibres, namely polypropylene, Kevlar and Courtelle (acrylic) for the Pluta and polyethylene and polyester for Interphako are presented. Figures 3.6(a,b) (plate 1) are two microinterferograms of a polypropylene fibre using the Pluta microscope with white light (a) totally duplicated and (b) differentially sheared (draw ratio = 5.2). Figures 3.7 (plate 2) are for the same fibre and immersion light. $n_L = 1.4800$ when monochromatic light of $\lambda = 546$ nm is used. Figure 3.8(a) (plate 3) shows a differentially sheared image of a highly oriented fibre, Kevlar 17, using white light; notice that the fringe shift is more than 18 orders of interference. Figure 3.8(b) (plate 3) shows totally duplicated images of Kevlar 17 fibre using the Pluta polarising interference microscope with monochromatic light of $\lambda = 546$ nm. Figure 3.9 shows a

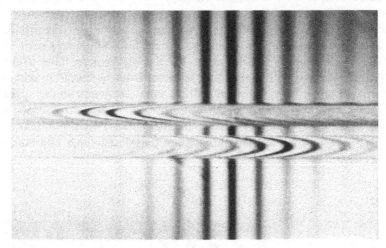

Figure 3.9 Totally duplicated images of a nylon 6 fibre (from an Egyptian manufacturer) using the Pluta microscope with white light. $n_L = 1.5080$ at $15\,°C$.

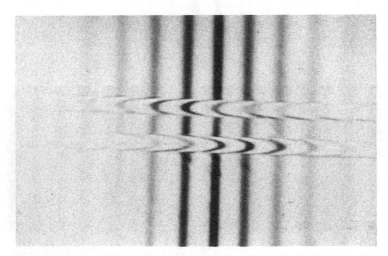

Figure 3.10 Totally duplicated images of a Courtelle (British Acrylic) fibre using the Pluta microscope with white light. $n_L = 1.5080$ at $15\,°C$.

totally duplicated image of nylon 6 fibres (from an Egyptian manufacturer) using the Pluta microscope with white light. Figure 3.10 shows totally duplicated images of a Courtelle (British Acrylic) fibre using the Pluta microscope with white light.

The Interphako interference microscope, as previously mentioned, is used for the measurement of refractive indices and birefringence of fibres. Figure 3.11 is a microinterferogram of a polyethylene fibre using the Interphako with white light vibrating perpendicular to the fibre axis, while figure 3.11(*b*) is for the same fibre with monochromatic light of $\lambda = 589.3$ nm vibrating parallel to the fibre axis. Figure 3.12 shows interferograms of a polyester fibre using the Interphako with monochromatic light of $\lambda = 589.3$ nm vibrating parallel and perpendicular to the fibre axis, respectively.

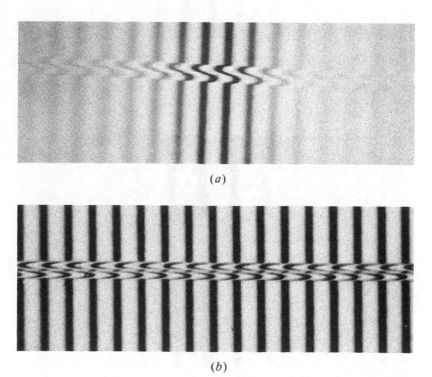

(*a*)

(*b*)

Figure 3.11 (*a*) Interferogram of polyethylene fibre using the Interphako interference microscope with white light vibrating perpendicular to the fibre axis. (*b*) Duplicated images of a polyethylene fibre using the Interphako with monochromatic light of wavelength $\lambda = 589.3$ nm vibrating parallel to the fibre axis.

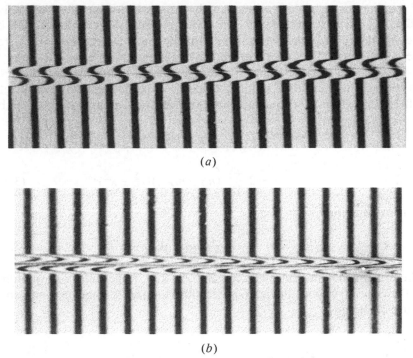

(*a*)

(*b*)

Figure 3.12 Interferograms of a polyester fibre using the Interphako interference microscope with monochromatic light of wavelength $\lambda = 589.3$ nm
vibrating parallel (*a*) and perpendicular (*b*) to the fibre axis, respectively.
(From Hamza 1986.)

3.3 Lateral birefringence of fibres

The refractive indices of a fibre may also vary across the fibre cross section
(see Morton and Hearle 1975). In this case fibres are not transversely
isotropic and these fibres acquire lateral birefringence. The mean refractive
indices $n_a^{\parallel}$ and $n_a^{\perp}$ were determined interferometrically, from one edge of an
unstretched viscose rayon model filament to another, by Faust (1956). In his
results, $n_a^{\parallel}$ was found constant to within ± 0.0001, while $n_a^{\perp}$ was found
greater at the edges than at the axis by 0.0015.

The lateral birefringence in Kevlar fibres was studied interferometrically
by Warner (1983) using a Leitz–Mach–Zehnder two-beam interference
microscope. Hamza *et al* (1989) gave an analysis of a system of two-beam
interference fringes crossing a cylindrical multilayer fibre.

Figure 3.13(*a*) represents a cross section of a cylindrical multilayer fibre of regular shape, having m layers and immersed in a liquid of refractive index n_L. The fibre cross section is in the (x, y) plane. The refractive index of the mth layer is n_m where $n_{m=1}$ is n_{skin} for the outer layer and $n_m = n_{core}$ for the inner layer. The radii of the fibre layers are given by $r_Q = (y_Q^2 + x^2)^{1/2}$, $Q = 1, 2, \ldots$. A parallel beam of plane polarised monochromatic light of wavelength λ is incident parallel to the y axis. The optical path length difference (OPLD) $\Delta\Gamma$ through the fibre and the immersion liquid is given by

$$\Delta\Gamma = \text{OPLD} = 2 \sum_{Q=1}^{m} (n_Q - n_{Q-1})y_Q$$

$$= 2 \sum_{Q=1}^{m} (n_Q - n_{Q-1})(r_Q^2 - x^2)^{1/2} \qquad (3.12)$$

and for equation (3.4)

$$\frac{\lambda}{2\,\Delta Z} Z = \sum_{Q=1}^{Q=m} (n_Q - n_{Q-1})(r_Q^2 - x^2)^{1/2} \qquad (3.13)$$

where ΔZ is the interfringe spacing and Z is the fringe shift corresponding to the value of x along the radius of the fibre on the (Z, x) plane as shown in figure 3.13(b).

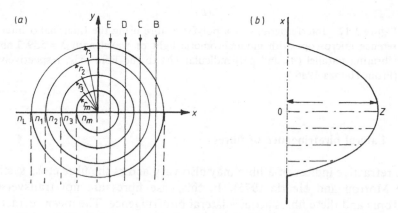

Figure 3.13 (a) A cross section in a cylindrical multilayer fibre. (b) The fringe shift across a cylindrical fibre.

The radial birefringence of a multilayer fibre of regular cross section is given by

$$Z^{\parallel} - Z^{\perp} = \frac{2\Delta Z}{\lambda} \sum_{Q=1}^{m} (\Delta n_Q - n_{Q-1})(r_Q^2 - x^2)^{1/2} \qquad (3.14)$$

where Δn_Q is the birefringence of the Qth layer. This fringe shift $Z^{\parallel} - Z^{\perp}$ is equal to the fringe shift in the case of a non-duplicated image using a polarising interference microscope, e.g. the Pluta microscope.

For a fibre having two layers, skin and core, the radial birefringence is obtained by substituting in equation (3.14) for $Q = 2$.

For a multilayer fibre of irregular cross section equation (3.8) is generalised leading to

$$F = \frac{\Delta Z}{\lambda} \sum_{Q=1}^{m} A_Q(n_Q - n_{Q-1}) \tag{3.15}$$

where

$$F = \int_{\alpha}^{\beta} Z \, dx \quad \text{and} \quad A = \int_{\alpha}^{\beta} t_Q \, dx.$$

This is reached by integrating the equation of a multilayer fibre of regular cross section over the region $\alpha \leqslant x \leqslant \beta$, since $t_Q = y_Q$ gives the area under the fringe shift.

Recently Pluta (1987) reported that a cylindrical birefringent fibre oriented diagonally between two crossed polarisers and immersed in a liquid, when normally transilluminated by monochromatic light and observed through a polarising microscope fitted with a condenser slit diaphragm, produces interference patterns in the exit pupil of the microscope objective. These patterns manifest themselves as optical Fourier transforms. They offer a new technique for measuring the mean birefringence of cylindrical fibres. Moreover, the spectral dispersion can be readily determined.

3.4 Applications of two-beam interferometric methods to optical fibres

Table 3.1 summarises the investigators who have applied two-beam interference microscopy to the study of optical fibre properties, namely the index profile of step-index and graded-index fibres, and their adopted technique. The table is compiled from surveys by Ghatak and Thyagarajan (1980), Marcuse and Presby (1980) and Okoshi (1982).

3.4.1 The interferometric slab method
A thin slab is cut out of the fibre of thickness 0.1–0.5 mm. The end faces are polished as the thickness of the slab should remain constant over the entire slab area to within a fraction of the wavelength of light. To measure the index profile of a fibre, a slab is placed in one arm of an interference microscope as shown in figure 3.14(a), and a homogeneous reference slab, with a refractive index n_2, is placed in the second arm of the interference microscope (figure 3.14(b)). If the fibre slab and reference slab are identical and the two mirrors are slightly inclined, straight-line, equally spaced fringes with two-beam intensity distribution are formed. When the fibre

Table 3.1 Interferometric determination of the optical properties of optical fibres.

Authors	Method	Results
Rawson and Murray (1973)	Interference between light reflected at both ends of the fibre	Determination of graded-index fibre parameters, C_4 and C_6, in $n^2(r) = n_0^2(1 - \delta^2 r^2 + C_4 \delta^4 r^4 + C_6 \delta^6 r^6 + \cdots)$
Martin (1974)	Interferometric slab method using a Michelson-type interference microscope	Index profile
Presby and Brown (1974)	Interferometric slab method	Graded-index profile, accuracy in index data to a few parts in 10^4 and a spatial resolution of 2 μm
Cherin et al (1974)	Interferometric slab method using a Mach–Zehnder system (Leitz interference microscope)	Refractive index measurement of three Corning multimode optical fibres
Burrus and Standley (1974)	Interferometric slab method	Viewing refractive index profiles and small-scale inhomogeneities in glass optical fibres
Burrus et al (1973)	Interferometric slab method	Refractive index profiles of some low-loss multimode optical fibres
Stone and Burrus (1975)	Interferometric slab method	Focusing effects in interferometric analysis of graded-index optical fibres
Presby and Kaminow (1976)	Interferometric slab method	Measured $dn/d\lambda$ for $0.5 < \lambda < 1.9$ μm with accuracy of 1 part in 10^5
Wonsiewicz et al (1976)	Interferometric slab method	Quick determination of index profiles by machine aided method for the interpretation of interferograms

Presby *et al* (1978)	Interferometric slab method using a two-beam single-pass interference microscope	Automatic index profiling
Shiraishi *et al* (1975)	Mach–Zehnder with light passing perpendicular to fibre axis	Index profile of graded-index fibres
Marhic *et al* (1975)	Two-beam transverse interference microscopy	Analytical expressions for OPLD for graded-index fibres with quadric index profile
Saunders and Gardner (1977)	Two-beam transverse interference microscopy	Index profile of graded-index. Determination of Δ and α of a fibre having a power law profile
Iga and Kokubun (1977, 1978)	Two-beam interference with light incident perpendicular to fibre axis	Graded-index profile, considering effect of refraction of the ray as it passes through the fibre
Kokubun and Iga (1977, 1978)	Two-beam interference with light incident perpendicular to fibre axis	They derived successive approximation formulae for calculating index profile
Iga *et al* (1976)	Differential interferometry (shearing). One beam is laterally shifted by a shearing device in a Mach–Zehnder interferometer	Measurement of index distribution of focusing fibres
Boggs *et al* (1979)	Transverse profile automated with computer controlled video analysis	Index profile of graded-index
Presby *et al* (1979)	Rapid automatic index profiling of whole fibre samples	Index profile of graded-index

slab under test is introduced in one arm of the interferometer, a fringe system appears as shown in figure 3.14(c). The fringe shift $S(x, y)$ depends upon its position in the fibre core. It corresponds to the relative phase difference ψ between the phase retardation in the fibre slab and the reference slab, where

$$\psi = \frac{2\pi}{\lambda}(n(x, y) - n_2)t \qquad (3.16)$$

$$\frac{2\pi}{D} = \frac{\psi}{S(x, y)} \qquad (3.17)$$

D being the fringe spacing and

$$(n(x, y) - n_2) = \frac{\lambda}{2\pi}\frac{\psi}{t} \qquad (3.18)$$

$$n(x, y) = n_2 + \frac{\lambda}{2\pi}\frac{2\pi S(x, y)}{Dt} = n_2 + \frac{\lambda S(x, y)}{Dt}. \qquad (3.19)$$

The fringe shift can be measured by recording the interferogram on a photographic plate and using a travelling microscope.

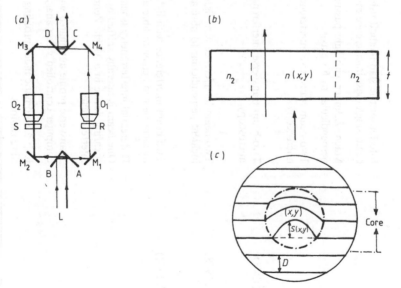

Figure 3.14　(a) A two-beam single-pass interference microscope. L is the incident light, M_1, M_2, M_3 and M_4 are mirrors, S is the slab, R is the reference slab, O_1 and O_2 are microscope objectives. A, B, C and D are semi-transparent mirrors.

(b) A slab of thickness t for a graded-index core with a clad of refractive index n_2.

(c) Interferogram in which the fringe shift $S(x, y)$ in the core region as a function of point position (x, y) is shown.

Plate 1

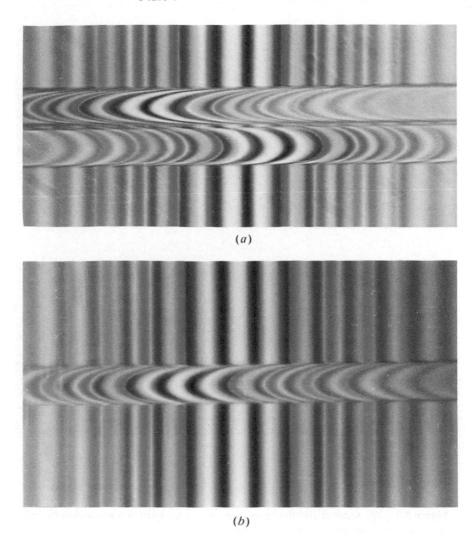

(*a*)

(*b*)

Figure 3.6 Totally duplicated (*a*) and differentially sheared (*b*) images of a polypropylene fibre (draw ratio = 5.2) using the Pluta polarising interference microscope with white light. $n_L = 1.4800$ at 17 °C.

Plate 2

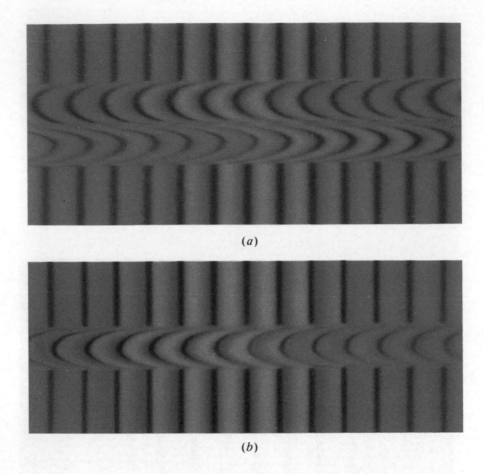

(a)

(b)

Figure 3.7 The same microinterferograms shown in figure 3.6 when using mono-chromatic light of wavelength $\lambda = 546$ nm.

Plate 3

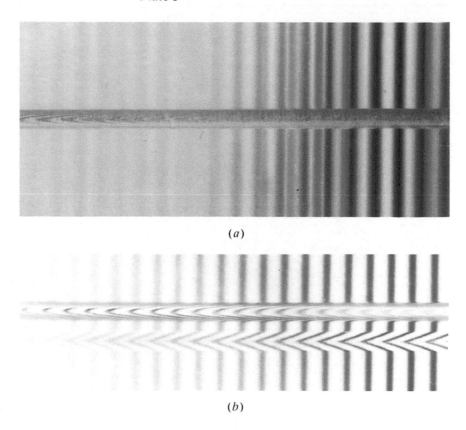

(*a*)

(*b*)

Figure 3.8 (*a*) Differentially sheared image of a highly oriented fibre of Kevlar 17. It is clear that the fringe shift is more than 18 orders of interference. (*b*) Totally duplicated images of a Kevlar 17 sample using the Pluta microscope with monochromatic light of wavelength $\lambda = 546$ nm.

Equation (3.19) is applicable when a transmission-type interference microscope is used. The ray passes through the sample once. For a Michelson-type interference microscope, shown in figure 3.15 (Cherin 1983), where the ray passes through the slab twice, the expression for $n(x, y)$ is as follows:

$$\Delta n(x, y) = \frac{S(x, y)}{D} \frac{\lambda}{2t}. \tag{3.20}$$

In this system, the spatial resolution and accuracy are $0.7\,\mu$m and $\pm 5 \times 10^{-4}$, respectively (Martin 1974). In practice the accuracy of the interferometric slab method is limited by the accuracy with which the slab thickness is measured and the degree of smoothness and parallelism of the slab end faces. The method is destructive to the fibre under test and requires time-consuming fibre sample preparation. Major sources of error are the effect of ray bending, especially when the sample is thick, and the roughness of the slab surfaces.

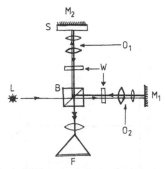

Figure 3.15 Set-up for measuring the index profile of fibres using a Michelson-type interferometer microscope. L is the light source, B the beam splitter, O_1 and O_2 microscope objectives, M_1 and M_2 mirrors, S is the slab and W the wavefront tilting device.

3.4.2 *Index profile of optical fibres from their transverse interference patterns*

Two-beam interference microscopy has been applied to fibres with light incident perpendicular to the fibre axis. The sample is introduced in one of the optical paths of a Mach–Zehnder interferometer, the optical arrangement being as shown in figure 3.16. The ray passes through the fibre normally. The fibre is immersed in a matching liquid whose refractive index n_L is nearly equal to that at the fibre surface n_{clad}. In such a case, the ray is considered to pass through the fibre almost straight and hence the phase-shift in the fibre can be worked out. Figure 3.17 shows the fibre immersed in a matching liquid (Shiraishi *et al* 1975). Assuming that the fibre is axially symmetrical and that the rays travel straight through the fibre the fringe shift is as shown in figure 3.18, the excess phaseshift being given in terms of the coordinate x as

$$Q(x) = 2k \int_x^R \frac{\Delta n(r) r \, dr}{(r^2 - x^2)^{1/2}} \tag{3.21}$$

where $\Delta n(r) = n(r) - n_L$, R is the radius of the cladding and

$$Q(x) = 2\pi \frac{d(x)}{D}$$

where D is the interfringe spacing and d the distance the chosen point moves. Therefore

$$d(x) = \frac{2D}{\lambda} \int_x^R \frac{\Delta n(r) \, r \, dr}{(r^2 - x^2)^{1/2}}. \tag{3.22}$$

This equation has the form of Abels' integral and can readily be solved by Abel inversion which is dealt with in Chapter 9.

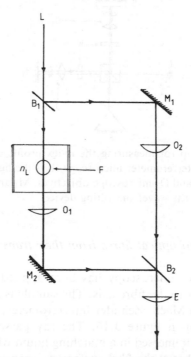

Figure 3.16 Mach–Zehnder interferometer transverse interference. L, light source; M_1 and M_2, opaque mirrors; O_1 and O_2, microscope objectives; B_1 and B_2, beam splitters; E, an eye-piece and F, fibre immersed in a matching-liquid bath.

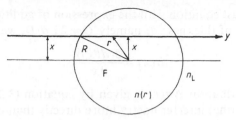

Figure 3.17   The fibre immersed in a matching liquid. F is the fibre, $n(r)$ is its refractive index and n_L that of the matching liquid.

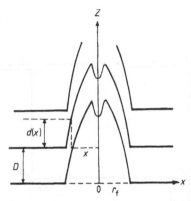

Figure 3.18   An interferogram across a fibre of radius r_f; Z is the fibre axis.

3.4.3   Differential interferometry

Iga *et al* (1976, 1978) modified the Mach–Zehnder interferometer by introducing a shearing device built in the interference microscope. Figure 3.19 shows the passage of light in the interferometer. One observes the interference pattern resulting from two beams which both pass through the fibre. One beam is laterally shifted by a small distance s, and what appears as the fringe shift is the phase difference between the two rays passing through the fibre at x and $x + s$, shown in figure 3.20.

$$Q_s = Q(x + s) - Q(x)$$

and for a small value of s,

$$Q_s = \lim_{s \to 0} \frac{Q(x+s) - Q(x)}{s} \, s = \frac{\mathrm{d}(Q(x))}{\mathrm{d}x} \, s. \tag{3.23}$$

Therefore, the fringe shift is

$$d_s(x) = \frac{D}{2\pi} \frac{\mathrm{d}(Q(x))}{\mathrm{d}x} \, s. \tag{3.24}$$

Comparing the last equation with the expression of additional phaseshift in a dielectric plate of thickness d, namely $Q = 2\pi\, d/D$, we have

$$d_s(x) = \frac{\mathrm{d}(d(x))}{\mathrm{d}x}\, s.$$

The index distribution $\Delta n(r)$ is given by equation (3.25) in this case of differential (shearing) interferometry more directly than by equation (3.22)

$$\Delta n(r) = -\frac{\lambda}{\pi D s} \int_r^R d_s(x) \frac{\mathrm{d}x}{(x^2 - r^2)^{1/2}}. \tag{3.25}$$

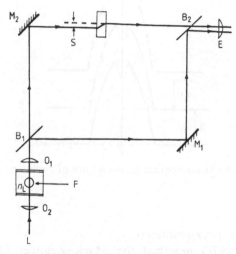

Figure 3.19 A shearing Mach–Zehnder interferometer for differential interferometry (the components are as in figure 3.16 and S is the shearing device).

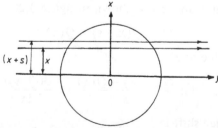

Figure 3.20 Two rays contribute to the formation of a differential interferogram resulting from shearing.

References

Barakat N 1971 *Textile Res. J.* **41** 167
Blakey P R, Montgomery D E and Sumner H M 1970 *J. Textile Inst.* **61** 234
Boggs L, Presby H M and Marcuse D 1979 *Bell Syst. Tech. J.* **58** 867
Bunn C W and Daubeny P 1954 *Trans. Faraday Soc.* **50** 1173
Burrus C A, Chinnock E L, Gloge D, Holden W S, Li T, Standley R D and Keck D B 1973 *Proc. IEEE* **61** 1498
Burrus C A and Standley R D 1974 *Appl. Opt.* **13** 2365
Cherin A H 1983 *An Introduction to Optical Fibres* (New York: McGraw-Hill)
Cherin A H, Cohen L Q, Holden W S, Burrus C A and Kaiser P 1974 *Appl. Opt.* **13** 2359
Denbigh K G 1940 *Trans. Faraday Soc.* **36** 936
Dorau K and Pluta M 1981a *Przeglad Włókienniczy* **35** 70
—— 1981b *Przeglad Włókienniczy* **35** 128
Faust R C 1956 *Q. J. Microsc. Sci.* **97** 569
Ghatak A and Thyagarajan 1980 *Progress in Optics* vol XVIII ed. E Wolf (Amsterdam: North-Holland) pp100–9
Hamza A A 1980 *Textile Res. J.* **50** 731
—— 1986 *J. Microsc.* **142** 35
Hamza A A and Abd El-Kader H I 1986 *Phys. Ed.* **21** 244
Hamza A A and El-Dessouki T 1987 *Textile Res. J.* **57** 508
Hamza A A and El-Farahaty K A 1986 *Textile Res. J.* **56** 580
Hamza A A, Fouda I M and El-Farahaty K A 1986 *Int. J. Polym. Mater.* **11** 169
Hamza A A, Kabeel M A and Shahin M M 1989 *Textile Res. J.* in press
Hamza A A and Sikorski J 1978 *J. Microsc.* **113** 15
Iga K and Kokubun Y 1977 *Tech. Digest Int. Conf., IOOC Tokyo* p403
—— 1978 *Appl. Opt.* **17** 1972
Iga K, Kokubun Y and Yamamoto N 1976 *Record of Natl. Symp. Light Radio Waves, IECE Japan* paper S3-1
—— 1978 *Papers of Technical Group IECE Japan* no OQE 76–80
Kokubun Y and Iga K 1977 *Trans. IEEC Japan* **E60** 702
—— 1978 *Trans. IECE Japan* **E61** 184
McKee A and Woods H J 1967 *J. R. Microsc. Soc.* **87** 185
McLean J H 1971 *Textile Res. J.* **41** 90
Marcuse D and Presby H 1980 *Proc. IEEE* **68** 6
Marhic M E, Ho P S and Epstein M 1975 *Appl. Phys. Lett.* **26** 574
Martin W E 1974 *Appl. Opt.* **13** 2112
Morton W E and Hearle J W S 1975 *Physical Properties of Textile Fibres* (London: The Textile Institute) pp573–8
Northolt M G 1974 *Europ. Polym. J.* **10** 799
Okoshi T 1982 *Optical Fibres (London: Academic)*
Pluta M 1965 *Przeglad Włókienniczy* **19** 261
—— 1971 *Opt. Acta* **18** 661
—— 1972 *J. Microsc.* **96** 309
—— 1982 *Mikroskopia Optyczna* (Warszawa: Państwowe Wydawnictwo Naukowe) (in Polish)

Pluta M 1987 *J. Mod. Opt.* **34** 1451
Presby H M and Brown W L 1974 *Appl. Phys. Lett.* **24** 511
Presby H M and Kaminow I P 1976 *Rev. Sci. Instrum.* **47** 348
Presby H M, Marcuse D and Astle H 1978 *Appl. Opt.* **14** 2209
Presby H M, Marcuse D, Boggs L and Astle H 1979 *Bell Syst. Tech. J.* **58** 883
Rawson E G and Murray R G 1973 *IEEE J. Quantum Electron.* **QE-9** 1114
Saunders M J and Gardner W B 1977 *Appl. Opt.* **16** 2369
Shiraishi S, Tanaka G, Suzuki S and Kurosaki S 1975 *Record of Natl. Conv., IECE Japan* **4** 239, paper 891
Sikorski J 1984 *Proc. R. Microsc. Soc.* **19** 28 (Book review)
Simmens S C 1958 *Nature* **181** 1260
Sokkar T Z N and Shahin M M 1985 *Textile Res. J.* **55** 139
Stone J and Burrus C A 1975 *Appl. Opt.* **14** 151
Warner S B 1983 *Macromolecules* **16** 1546
Wilkes J M 1985 *Textile Res. J.* **55** 712
Wonsiewicz B C, French W G, Lazay P D and Simpson J R 1976 *Appl. Opt.* **15** 1048
Żurek W and Zakrzewski S 1983 *J. Appl. Polym. Sci.* **28** 1277

4 Multiple-beam Interferometry Applied to Fibrous Materials

4.1 Formation and application of multiple-beam interference fringes to fibres

Multiple-beam interference methods have been developed by Tolansky since 1944. The types of multiple-beam interference fringes normally applied to fibres are (i) multiple-beam Fizeau fringes in transmission, (ii) multiple-beam Fizeau fringes at reflection and (iii) multiple-beam fringes of equal chromatic order both in transmission and at reflection.

It is helpful when dealing with multiple-beam Fizeau fringes in transmission and at reflection to present the system of multiple-beam interference fringes formed by a plane parallel doubly silvered film of constant thickness. The reason lies in the fact that there is a close resemblence between the properties of the two types of fringe systems. The Fabry–Perot interferometer is an example of the application of cases (i) and (ii), and is in fact an idealised case of multiple-beam Fizeau fringes. Multiple-beam fringes in transmission and at reflection have different applications.

4.1.1 The case of multiple-beam interference fringe systems formed by a plane parallel doubly silvered thin film of constant thickness

Figure 4.1 represents parallel monochromatic light incident on two plane parallel plates silvered on the inner surface. The incident beam makes an angle θ with the normal to the plates.

The expression for the resultant of the multiple-reflected beam in transmission formed by a film of constant thickness t and refractive index n

coated with a reflecting metallic layer is given by

$$R_T = T_1 T_2 \exp[i(\omega t + \gamma_1 + \gamma_2)] + T_1 T_2 R_2 R_3 \exp\{i[(\omega t + \gamma_1 + \gamma_2) + \Delta]\}$$
$$+ T_1 T_2 R_2^2 R_3^2 \exp\{i[(\omega t + \gamma_1 + \gamma_2) + 2\Delta]\} + \cdots. \quad (4.1)$$

The phase properties of the metallic coating are defined as follows: β_1, change of phase at reflection at the glass/metallic layer interface for the upper component facing the incident light; β_2 and β_3, change of phase at reflection at the medium/metallic layer interface, for the upper and lower component, respectively; γ_1 and γ_2, change of phase in transmission through the metallic layers of the upper and the lower components, respectively. R_1^2 and R_2^2 are the fractions of light intensity reflected at the glass/metallic layer and medium/metallic layer interfaces, for the upper component. R_3^2 is the fraction of light intensity reflected at the medium/layer interface for the second component. T_1^2 and T_2^2 are the fractions of light intensity transmitted through the metallic layers of the upper and lower components, respectively. Δ is the constant phase difference between any two successive rays and ω is the frequency.

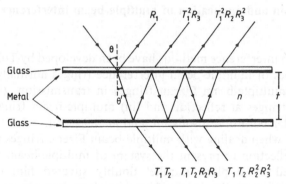

Figure 4.1 Multiple-beam interference in a thin film, of constant thickness t, in transmission and at reflection.

From equation (4.1) one can get

$$R_T = T_1 T_2 \left(\frac{1}{1 - R_2 R_3 \exp(i\Delta)} \right) \exp[i(\omega t + \gamma_1 + \gamma_2)] \quad (4.2)$$

$$R_T = T_1 T_2 \left(\frac{1 - R_2 R_3 \exp(-i\Delta)}{[1 - R_2 R_3 \exp(i\Delta)][1 - R_2 R_3 \exp(-i\Delta)]} \right)$$
$$\times \exp[i(\omega t + \gamma_1 + \gamma_2)]$$

$$= T_1 T_2 \left(\frac{1 - R_2 R_3 \cos \Delta + i R_2 R_3 \sin \Delta}{1 + R_2^2 R_3^2 - R_2 R_3 \exp(-i\Delta) - R_2 R_3 \exp(i\Delta)} \right)$$
$$\times \exp[i(\omega t + \gamma_1 + \gamma_2)].$$

$$R_T = T_1 T_2 \left(\frac{1 - R_2 R_3 \cos \Delta + iR_2 R_3 \sin \Delta}{1 + R_2^2 R_3^2 - R_2 R_3 [\exp(i\Delta) + \exp(-i\Delta)]} \right)$$

$$\times \exp[i(\omega t + \gamma_1 + \gamma_2)]$$

$$= T_1 T_2 \left(\frac{1 - R_2 R_3 \cos \Delta + iR_2 R_3 \sin \Delta}{1 - 2R_2 R_3 \cos \Delta + R_2^2 R_3^2} \right) \exp[i(\omega t + \gamma_1 + \gamma_2)]$$

$$R_T = A_T \exp[i(\omega t + \gamma_1 + \gamma_2 + \Delta_T)]. \tag{4.3}$$

A_T is the amplitude of the resultant and Δ_T is its phase difference with respect to the first transmitted beam.

$$\Delta = \frac{2\pi}{\lambda} (2nt \cos \theta) + \beta_3 + \beta_2$$

$$= \delta + \beta_3 + \beta_2$$

therefore $I_T = A_T^2$

$$= T_1^2 T_2^2 \left(\frac{(1 - R_2 R_3 \cos \Delta + iR_2 R_3 \sin \Delta)(1 - R_2 R_3 \cos \Delta - iR_2 R_3 \sin \Delta)}{(1 - 2R_2 R_3 \cos \Delta + R_2^2 R_3^2)^2} \right)$$

$$= T_1^2 T_2^2 \frac{1 - 2R_2 R_3 \cos \Delta + R_2^2 R_3^2}{(1 - 2R_2 \cos \Delta + R_2^2 R_3^2)^2}$$

$$= \frac{T_1^2 T_2^2}{1 - 2R_2 R_3 \cos \Delta + R_2^2 R_3^2} \tag{4.4}$$

$$\tan \Delta_T = \frac{R_2 R_3 \sin \Delta}{1 - R_2 R_3 \cos \Delta}.$$

4.1.2 The intensity distribution of multiple-beam Fizeau fringes at reflection

The resultant of multiple-beam Fizeau fringes at reflection is given by

$$R_R = R_1 \exp[i(\omega t + \beta_1)] + T_1^2 R_3 \exp[i(\omega t + 2\gamma_1 + \beta_3 + \delta)]$$
$$+ T_1^2 R_2 R_3^2 \exp[i(\omega t + 2\gamma_1 + \beta_2 + 2\beta_3 + 2\delta)] + \cdots.$$

Putting $\Delta = \delta + \beta_2 + \beta_3$ and $F = 2\gamma_1 - \beta_1 - \beta_2$ we get

$$R_R = R_1 \exp[i(\omega t + \beta_1)] + T_1^2 R_3 \exp[i(\omega t + \beta_1)]\exp[i(F + \Delta)]$$
$$+ T_1^2 R_2 R_3^2 \exp[i(\omega t + \beta_1)]\exp[i(F + 2\Delta)] + \cdots$$

$$= \{R_1 + T_1^2 R_3 \exp[i(F + \Delta)][1 + R_2 R_3 \exp(i\Delta) + R_2^2 R_3^2 \exp(i2\Delta) + \cdots]\}$$
$$\times \exp[i(\omega t + \beta_1)]$$

$$= \left[R_1 + T_1^2 R_3 \exp[i(F + \Delta)] \right.$$

$$\left. \times \left(\frac{1 - R_2 R_3 \exp(-i\Delta)}{[1 - R_2 R_3 \exp(i\Delta)][1 - R_2 R_3 \exp(-i\Delta)]} \right) \right] \exp[i(\omega t + \beta_1)]$$

$$= \left[R_1 + T_1^2 R_3 \left(\frac{\exp[i(F + \Delta)] - R_2 R_3 \exp(iF)}{1 - 2R_2 R_3 \cos \Delta + R_2^2 R_3^2} \right) \right] \exp[i(\omega t + \beta_1)].$$

$$R_R = \left[R_1 + T_1^2 R_3 \left(\frac{\cos(F + \Delta) - R_2 R_3 \cos F + i \sin(F + \Delta) - i R_2 R_3 \sin F}{(1 - 2R_2 R_3 \cos \Delta + R_2^2 R_3^2)} \right) \right]$$

$$\times \exp[i(\omega t + \beta_1)]$$

$$= \left[R_1 + T_1^2 R_3 \left(\frac{\cos(F + \Delta) - R_2 R_3 \cos F + i[\sin(F + \Delta) - R_2 R_3 \sin F]}{(1 - 2R_2 R_3 \cos \Delta + R_2^2 R_3^2)} \right) \right]$$

$$\times \exp[i(\omega t + \beta_1)]$$

$$= \left(R_1 + \frac{T_1^2 R_3 [\cos(F + \Delta) - R_2 R_3 \cos F]}{1 - 2R_2 R_3 \cos \Delta + R_2^2 R_3^2} \right.$$

$$\left. + i \frac{T_1^2 R_3 [\sin(F + \Delta) - R_2 R_3 \sin F]}{1 - 2R_2 R_3 \cos \Delta + R_2^2 R_3^2} \right) \exp[i(\omega t + \beta_1)].$$

$$I_R = \left(R_1 + \frac{T_1^2 R_3 [\cos(F + \Delta) - R_2 R_3 \cos F]}{1 - 2R_2 R_3 \cos \Delta + R_2^2 R_3^2} \right)^2$$

$$+ T_1^4 R_3^2 \left(\frac{\sin(F + \Delta) - R_2 R_3 \sin F}{1 - 2R_2 R_3 \cos \Delta + R_2^2 R_3^2} \right)^2$$

$$= R_1^2 + \frac{T_1^4 R_3^2 + 2T_1^2 R_1 R_3 \cos(F + \Delta) - 2T_1^2 R_1 R_2 R_3^2 \cos F}{1 - 2R_2 R_3 \cos \Delta + R_2^2 R_3^2}.$$

The final equation gives the intensity distribution I_R of the reflected system of fringes for any value of F.

(*a*) When $F = 2\pi m$

$$I_R = R_1^2 + \frac{T_1^4 R_3^2 + 2T_1^2 R_1 R_3 \cos \Delta - 2T_1^2 R_1 R_2 R_3^2}{1 - 2R_2 R_3 \cos \Delta + R_2^2 R_3^2}$$

$$= R_1^2 - \frac{T_1^2 R_1}{R_2} + \frac{T_1^4 R_3^2 - T_1^2 R_1 R_2 R_3^2 + (T_1^2 R_1 / R_2)}{1 - 2R_2 R_3 \cos \Delta + R_2^2 R_3^2}$$

$$= A - B + \frac{C}{1 - 2R_2 R_3 \cos \Delta + R_2^2 R_3^2}$$

where

$$A = R_1^2$$

$$B = T_1^2 R_1 / R_2$$

and

$$C = T_1^4 R_3^2 - T_1^2 R_1 R_2 R_3^2 + T_1^2 R_1 / R_2.$$

(*b*) When $F = (2m + 1)\pi$

$$I_R = R_1^2 + \frac{T_1^4 R_3^2 - 2T_1^2 R_1 R_3 \cos \Delta + 2T_1^2 R_1 R_2 R_3^2}{1 - 2R_2 R_3 \cos \Delta + R_2^2 R_3^2}$$

$$= R_1^2 + (T_1^2 R_1/R_2) + \frac{T_1^4 R_3^2 + T_1^2 R_1 R_2 R_3^2 - (T_1^2 R_1/R_2)}{1 - 2R_2 R_3 \cos \Delta + R_2^2 R_3^2}$$

$$= A + B - \frac{D}{1 - 2R_2 R_3 \cos \Delta + R_2^2 R_3^2}$$

where

$$D = (T_1^2 R_1/R_2) - (T_1^4 R_3^2 + T_1^2 R_1 R_2 R_3^2).$$

4.1.3   *Analysis of elements determining the shape of the intensity distribution (Mokhtar 1964)*

The three systems of interference formed by a Fabry–Perot interferometer are listed below.

(*a*) Multiple-beam reflected system characterised by sharp dark lines on a bright background.

(*b*) Multiple-beam transmitted system characterised by sharp bright lines on a dark background.

(*c*) Transmitted-like fringes, formed at reflection with an intensity distribution identical to the transmitted system but with higher peak intensity and background.

From the previous theoretical investigations it can be shown that the intensity distribution for any of the above mentioned three systems can be represented by a general expression of the form

$$I = A + B + \frac{C}{1 - 2R_2 R_3 \cos \Delta + R_2^2 R_3^2}.$$

For the transmitted system

$$A = B = 0$$
$$C = T_1^2 T_2^2.$$

This is the usual intensity distribution given by Airy summation where

$$I_{max} = \frac{T_1^2 T_2^2}{(1 - R_2 R_3)^2} \qquad \text{for } \Delta = 2\pi S, \; S = 0, 1, 2, \dots$$

and

$$I_{min} = \frac{T_1^2 T_2^2}{(1 + R_2 R_3)^2} \qquad \text{for } \Delta = (2S + 1)\pi, \; S = 0, 1, \dots.$$

For the reflected system

$$A = R_1^2$$
$$B = T_1^2 R_1/R_2$$
$$C = -[\mp (T_1^2 R_1/R_2) - R_1^4 R_3^2 \pm T_1^2 R_1 R_2 R_3^2].$$

It is clear that the intensity distribution is determined by the values of A, B and C. The contribution of A is a uniform intensity for all Δ. This is also the case for B if both B and A are positive, and the net result is an elevation of intensity for all values of Δ equal to the sum of the intensities corresponding to $A + B$. While if B is negative but still $|B|$ is less than A, then the net result is an elevation of intensities for all Δ equal to $A - B$. The last term of the general expression gives an intensity which varies with Δ, it is obviously an intensity distribution of a Fabry–Perot system represented by Airy's summation having I_{max} and I_{min} at $\Delta = 2S\pi$ and $(2S + 1)\pi$, respectively. Now if C is positive, the transmitted system given by the last term will be elevated for all Δ by $A + B$ as a background intensity. Clearly if B is negative the difference is positive and this brings down the background to $A - |B|$. Now if C is negative while both A and B are positive and a plane mirror is placed on the Δ axis at $I = A + B$ the net result of the three terms will be sharp dark lines on a bright background which is the mirror image of the transmitted system given by the last term. In this case the absolute value of I_{max} contributed from the last term must be less than $A + B$. It is equal to $A + B - [C/(1 - R_2R_3)^2]$. This gives I_{min} for the multiple-beam reflected system which takes place at $\Delta = 2S\pi$. The net value of I_{max} is equal to $A + B - [C/(1 + R_2R_3)^2]$ which occurs at $\Delta = (2S + 1)\pi$. As has been mentioned, $A + B$ gives a constant value of intensity for all Δ, and the third term

$$\frac{C}{(1 - 2R_2R_3 \cos \Delta + R_2^2 R_3^2)}$$

gives a transmitted system, which in this case is subtracted from $A + B$, with intensity

$$I_{max} = \frac{C}{(1 - R_2R_3)^2} \qquad \text{at } \Delta = (2S + 1)\pi$$

and

$$I_{min} = \frac{C}{(1 + R_2R_3)^2} \qquad \text{at } \Delta = 2S\pi.$$

This is presented in figure 4.2. The net result being

$$I_{max} = R_1^2 + (T_1^2 R_1/R_2) - \frac{(T_1^2 R_1/R_2) - (T_1^4 R_3^2 + T_1^2 R_1 R_2 R_3^2)}{(1 + R_2R_3)^2}$$

$$= \left(R_1 + \frac{T_1^2 R_3}{(1 + R_2R_3)}\right)^2 \qquad \text{for } \Delta = (2S + 1)\pi$$

and

$$I_{min} = R_1^2 + (T_1^2 R_1/R_2) - \frac{(T_1^2 R_1/R_2) - (T_1^4 R_3^2 + T_1^2 R_1 R_2 R_3^2)}{(1 - R_2 R_3)^2}$$

$$= \left(R_1 - \frac{T_1^2 R_3}{(1 - R_2 R_3)} \right)^2 \qquad \text{for } \Delta = 2S\pi.$$

Now if the first beam is masked then $A - B = 0$ and the net result will be a system with a transmitted intensity distribution.

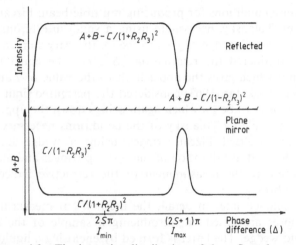

Figure 4.2   The intensity distribution of the reflected system.

4.1.4   *Multiple-beam Fizeau fringes formed by a silvered wedge*

Tolansky (1948) carried out the analysis for the conditions necessary to produce multiple-beam localised Fizeau fringes using a wedge interferometer. He pointed out that the Airy summation, strictly speaking, only holds for a parallel plate, but if certain critical conditions are fulfilled for a silvered air wedge, then a close approximation to the Airy summation can be achieved.

Tolansky pointed out the main distinguishing feature between beams forming multiple-beam fringes at infinity by using a plane parallel plate and those forming multiple-beam localised fringes. In the case of the silvered air wedge, the successively multiple-reflected beams are not behind each other in phase in exact arithmetic series, while in the case of a plane parallel plate the phase difference Δ between any two successive beams is independent of the order of the multiple-reflected beam as given by the equation

$$\Delta = \frac{2\pi}{\lambda} (2nt \cos \theta) + \beta_2 + \beta_3.$$

But for a silvered air wedge, the phase lag of the multiple-reflected beams from the arithmetic series is equal to

$$\tfrac{4}{3}\pi S^3 \varepsilon^2 N$$

where ε is the angle of the wedge, S is the order of the beam and N is the order of the interference. The incidence being normal the path lag is equal to

$$\frac{\lambda}{2\pi}\left(\frac{4}{3}\pi S^3 \varepsilon^2 \frac{2t}{\lambda}\right) = \frac{4}{3} S^3 \varepsilon^2 t.$$

The optimum conditions for producing multiple-beam Fizeau fringes, as determined by Tolansky, necessitate using a small interferometric gap of thickness t and small wedge angle ε to secure the Airy sum conditions.

Tolansky considered the retardation $\tfrac{4}{3} S^3 \varepsilon^2 t$ to be equal to $\lambda/2$ as a permitted limit which gives the upper limit to the value of t and ε.

Barakat and Mokhtar (1963) considered the permitted limit which gives maximum intensity to be $\tfrac{3}{8}\lambda$ which again brings down the upper limit of t.

The analysis made by Tolansky of the conditions necessary to produce multiple-beam localised Fizeau fringes using a wedge interferometer extended to the field of utilisation of these sharp fringes and made possible their application to the measurement of the refractive indices and bire-fringence of fibres.

As it is explained later in detail, the methods of measuring refractive indices of fibres are based on introducing a sample of the fibre into a silvered liquid wedge. The latter is formed by enclosing a liquid of suitable refractive index between two silvered optical flats forming a wedge of small angle, and at the same time introducing the fibre perpendicular to the edge of the wedge.

As previously mentioned both the interferometric gap nt and the wedge angle ε should be kept small to bring down the phase lag between successive beams.

4.1.5 Fringes of equal chromatic order

The interference taking place when a parallel beam of monochromatic light is incident on a wedge with highly reflecting surfaces has been dealt with. In addition to the well known plane of fringe localisation close to the wedge, i.e. the Feussner (1927) surface of localisation whose position is independent of the wavelength used, Brossel (1947) discovered the presence of an infinite number of other planes of localisation at a distance x from the Feussner surface given by $x = m\lambda/2\alpha^2$ for the case of normal incidence, α being the angle of the air wedge, λ the wavelength used and m takes the values $1, 2, 3, \dots$.

The dependence on the wavelength of the distance x of such planes of localisation is apparent. A variation of $d\lambda$ in λ produces a shift in the plane of localisation along the x axis equal to $(x/\lambda)\, d\lambda$.

The formation of fringes of equal chromatic order

Any point on the Feussner surface has a certain thickness t and interference takes place in transmission when the basic condition $N\lambda = 2t \cos\theta$ is satisfied for a certain wavelength λ and order N. The same condition holds for another wavelength λ_1, for the $(N+1)$th order, ... λ_m for the $(N+m)$th order. This applies whether the wavelengths are present separately or in a continuous spectrum. Because the Feussner surface is independent of the wavelength, these point fringes belonging to different wavelengths are superimposed on each other. They cannot be seen on the Feussner surface, except for very low values of N when they appear as coloured bands starting from the violet to the red for $N = 1$ and vanish due to overlap for higher orders. Now if the Feussner surface is projected on the slit of a spectrograph by an achromat, or the interferometer brought close to the slit, the slit selects a line on the surface, and, in general, the thickness t varies along this line. Considering the line to be an infinite number of these point fringes, the dispersive power of the spectrograph separates each set of superimposed points, to be seen on the spectral plane. Let us consider any two points on the line selected by the slit corresponding to t and $t + dt$. Then for the same order of interference, two point fringes are seen on the spectral plane at wavelengths λ and $\lambda + d\lambda$ where $t/\lambda = (t + dt)/(\lambda + d\lambda) = \text{constant} \times N$.

If the thickness t changes gradually over dt, a continuous curve for every order of interference results. In the case of a vertical step t changes abruptly and a discontinuity occurs. The family of white light interference fringes on the spectral plane whose order of interference is constant along every single member, are the fringes of equal chromatic order discovered by Tolansky in 1945.

The conditions for formation

If the monochromatic localised fringes on any of the principal Brossel planes are projected on a spectrograph slit and a white light source is used, the resulting white light fringes seen on the spectral plane are only in focus over a very limited area, depending upon the extension of localisation in space as well as the depth of focus of the projecting achromat.

Barakat (1957) pointed out that the associated white light fringes of equal chromatic order can only be formed in focus on the spectral plane if the position of the surface of localisation of the monochromatic system is independent of the wavelength used. Applying this conclusion to the monochromatic fringes localised on the Feussner surface of zero order (i.e. $m = 0$), the associated fringes of equal chromatic order obtained on the spectral plane are all in focus.

The shape of fringes of equal chromatic order

It is clear that the shape of the resulting fringes depends basically upon the way t varies along the line selected by the slit. If this line is taken to be the y axis, then in general t is a function of y, i.e. $t = f(y)$. The spectral plane is the (λ, y) plane, and thus the shape of fringes of equal chromatic order results directly from transforming the equation $t = f(y)$ from the (t, y) plane to the (λ, y) plane, obeying the relation $N\lambda = 2nt \cos \theta$ for the transmitted system and neglecting phase changes at reflection. The shape of the resulting fringes depends on the transforming relation. For the reflected system where the relation is $(N + \frac{1}{2})\lambda = 2nt \cos \theta$, the narrow dark lines have the same shape as the transmitted fringes. Two factors need to be considered: (i) the magnification of the projecting lens and the possible magnification of the spectrograph and (ii) the dispersion of the spectrograph. The effect of the magnification due to the projecting lens is only along the direction of the slit; it has no effect along the perpendicular direction—the λ axis—where the dispersive power of the instrument functions. A prism spectrograph or a diffraction grating could be used; in the first case, the dispersion D obeys Hartman's formula $\lambda = \lambda_0 + B/(D - D_0)$ where λ_0, B and D_0 are constants, while the grating, using the first order of diffraction, provides a linear dispersion $D = K\lambda$. Confining ourselves to linear dispersion, and substituting for λ and t in terms of D and y in the basic equation

$$N\lambda = 2n \cos \theta \; f\!\left(\frac{y}{m}\right)$$

$D = (2K/N)f(y/m)$ for any fringe, at normal incidence using an air medium. This is an equation of a family of fringes of decreasing magnification as N takes the values $1, 2, 3, \ldots$. The above relation shows that any fringe on the (D, y) plane is a magnified image of the section of the interferometer selected by the slit, and since the effect of the magnification is in general not the same along the D and Y axes, a distorted image is obtained. As an example, elliptic fringes result from a circular section. Using a prism spectrograph, another cause of distortion arises which is due to the non-linearity of the dispersion. In particular this is evident with fringes extending over a considerable range of wavelength, e.g. the bending of straight-line fringes resulting from a wedge.

The shape of fringes of equal chromatic order formed by an air wedge

If α is the angle of the air wedge, one of whose components is adjusted parallel to the plane of the slit, and ε the optical separation of the two components at the point of contact, then $(t - \varepsilon)/y = \tan \alpha$ is the equation for the section of the wedge selected by the slit.

$y = \cot \alpha \; (t - \varepsilon)$ is a linear function of t. Substituting in the basic equation, at normal incidence, for interference in transmission $N\lambda = 2t$

$$N\lambda = 2 \tan \alpha \; y + 2\varepsilon$$

therefore

$$y = \cot \alpha \left(\frac{N\lambda}{2} - \varepsilon\right)$$

$$= \frac{N \cot \alpha}{2} \left(\lambda - \frac{2\varepsilon}{N}\right). \tag{4.5}$$

Equation (4.5) represents a family of non-parallel straight lines, the slope of each member is $N/2 \cot \alpha$ where N is an integer. The family has a common point at $(0, -\varepsilon \cot \alpha)$ and as N increases the fringes approach the vertical direction of the y axis.

The case of Newton's rings
A lens and an optical flat, coated with semi-transparent highly reflecting silver layers, are brought into contact. The equation of a circle in the (t, y) plane is

$$[t - (R + \varepsilon)]^2 + y^2 = R^2$$

where R is the radius of curvature and ε is the optical separation at the point of contact. The origin is chosen such that $t = \varepsilon$ at $y = 0$. Transforming the previous equation to the (λ, y) plane

$$\left(\lambda - \frac{2(R + \varepsilon)}{N}\right)^2 \left(\frac{4R^2}{N^2}\right)^{-1} + \frac{y^2}{R^2} = 1.$$

This equation represents a family of ellipses whose centres are $(2(R + \varepsilon)/N, 0)$ and whose semi-major and semi-minor axes are R and $2R/N$, as N takes the positive integer values.

Since t is small compared with R, t^2 could be neglected. The resulting equation represents a family of parabolas, as reached by Tolansky. These fringes are convex to the violet.

Using a plano-concave lens with its curved surface as one of the components of the interferometer, the equation of the family of the fringes is

$$\left(\lambda + \frac{2(R - \varepsilon)}{N}\right)^2 \left(\frac{4R^2}{N^2}\right)^{-1} + \frac{y^2}{R^2} = 1$$

where ε is the sagitta of the curved surface of radius R, with respect to the optical flat. This is an equation of a family of ellipses whose centres are $(2\varepsilon/N, 0)$ and which are convex to the red. Thus for a hill the fringes of equal chromatic order are convex to the violet, while for a valley they are convex to the red. An easy way of relating the concavity of the fringes to the curved surface under investigation is to look for the image of the curved surface in the optical flat. Figure 4.3 shows the set-up used for the formation of fringes of equal chromatic order.

Figure 4.4 shows how fringes of equal chromatic order are formed from fringes of equal thickness. Elliptic fringes of equal chromatic order result from a lens-plate interferometer where systems of concentric circles of equal thickness are formed by monochromatic light from mercury spectral lines.

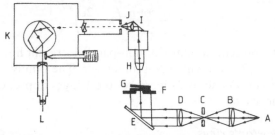

Figure 4.3 The set-up used for the formation of fringes of equal chromatic order in transmission. (From Barakat and El-Hennawi 1971.)

A, Pointolite white light source; B, condenser lens; C, iris diaphragm; D, collimating lens; E, 45° reflector; F, microscope stage; G, silver liquid wedge; H, low-power microscope objective; I, right-angle prism; J, projection lens; K, constant deviation prism spectrometer; L, to camera attached.

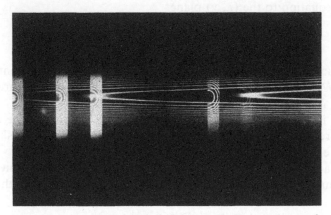

Figure 4.4 To show how fringes of equal chromatic order are formed from fringes of equal thickness.

4.2   Application of multiple-beam Fizeau fringes to the determination of refractive indices of fibres

In §4.1 the theory of formation, localisation and intensity distribution of multiple-beam Fizeau fringes and the fringes of equal chromatic order formed by a silvered air or liquid wedge were presented. In this section we are concerned with the application of these multiple-beam fringes to fibrous materials. It constitutes the theory of interferometric determination of indices of fibres.

4.2.1 Theory of Fizeau fringes applied to fibres with regular transverse sections

Mathematical expressions were derived by Barakat (1971) for the shape of multiple-beam Fizeau fringes crossing a fibre of circular transverse section immersed in a silvered liquid wedge. Formulae were obtained for calculating from the fringe shift the refractive indices of heterogeneous fibres with skin–core structure. A formula is deduced for a single-medium homogeneous fibre.

Theory

The following gives the mathematical equation of the family of Fizeau fringes across a fibre of perfectly circular cross section and having a core surrounded by a skin (idealised case). In interferometric investigation of fibres the fibre is introduced in a silvered liquid wedge, with the fibre axis adjusted perpendicular to the edge of the wedge. Let us assume that one of the two silvered optical flats just touches the circumference of the fibre. Figure 4.5 represents a cross section of a cylindrical fibre of radius r_f, having a core of refractive index n_c, radius r_c and a skin of refractive index n_s. It is immersed in a liquid wedge enclosing a liquid of refractive index n_L. A parallel beam of monochromatic light of wavelength λ is incident along AB and CD, normal to the lower component of the interferometric gap, the wedge angle ε being small. The thickness of the interferometric gap is represented by t. The fibre axis is chosen as the Z axis, and the edge of the wedge is parallel to the X axis. Two regions are to be considered (i) $X^2 + Y^2 = r_c^2$ and (ii) $X^2 + Y^2 = r_f^2$ where $r_c \leqslant X \leqslant r_f$. The shape of the fringes in the (Z, X) plane, which is the plane of the interferogram, is derived as follows. For the optical path length (OPL) of the ray AB crossing the fibre in the region $0 \leqslant X \leqslant r_c$

$$\text{OPL} = (t - 2Y_2)n_L + 2(Y_2 - Y_1)n_s + 2Y_1 n_c. \tag{4.6}$$

On a fringe of interference order N

$$N\lambda = 2n_L t + 4Y_2(n_s - n_L) + 4Y_1(n_c - n_s) \tag{4.7}$$

where $t = Z \tan \varepsilon$, ε being the wedge angle and the projected edge of the wedge being the Z origin. Rearranging (4.7)

$$N\lambda - 2n_L t = 4Y_2(n_s - n_L) + 4Y_1(n_c - n_s)$$

$$2n_L \tan \varepsilon \left(\frac{N\lambda}{2n_L \tan \varepsilon} - Z \right) = 4Y_2(n_s - n_L) + 4Y_1(n_c - n_s). \tag{4.8}$$

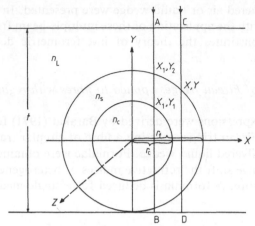

Figure 4.5 Cross section of a cylindrical fibre of radius r_f, core of refractive index n_c, radius r_c, and skin of index n_s immersed in a silvered liquid wedge. The refractive index of the enclosed liquid is n_L. (From Barakat 1971.)

Transforming the origin to the point $(N\lambda/2n_L \tan \varepsilon, 0)$ on the (Z, X) plane we have

$$2n_L \tan \varepsilon\, Z = 4Y_2(n_s - n_L) + 4Y_1(n_c - n_s)$$
$$= 4(n_s - n_L)(r_f^2 - X^2)^{1/2} + 4(n_c - n_s)(r_c^2 - X^2)^{1/2}. \quad (4.9)$$

Here Z measures the fringe shift of the fringe of the nth order in the fibre region from its position in the liquid region, the Z direction being towards the wedge apex. Let ΔZ represent the linear separation between any two consecutive fringes in the liquid region. The wedge angle ε is related to ΔZ by $\tan \varepsilon = \lambda/2n_L \Delta Z$

$$\frac{Z}{\Delta Z}\frac{\lambda}{4} = (n_s - n_L)(r_f^2 - X^2)^{1/2} + (n_c - n_s)(r_c^2 - X^2)^{1/2}. \quad (4.10)$$

At $X = 0$

$$\frac{Z}{\Delta Z}\frac{\lambda}{2} = (n_s - n_L)2r_f + (n_c - n_s)2r_c$$

$$= (n_s - n_L)t_f + (n_c - n_s)t_c$$

where $t_f = 2r_f$ and $t_c = 2r_c$ and

$$\frac{Z}{\Delta Z}\frac{\lambda}{2} = n_s t_s + n_c t_c - n_L t_f \quad (4.11)$$

where $t_s = (t_f - t_c)$.

$$\frac{Z}{\Delta Z}\frac{\lambda}{2} = (n_a - n_L)t_f \tag{4.12}$$

where $n_a = n_c t_c/t_f + n_s t_s/t_f$.

In order to apply these relations to actual measurements $Z/\Delta Z$ is measured and knowing t_f and n_L, n_a can be calculated. Then by determining n_s using the Becke-line method and t_s/t_f by, for example, differential staining, n_c can be calculated. By determining the value of the fringe shift $Z^{\parallel}$ as a fraction of an order separation ΔZ between straight-line consecutive fringes in the liquid region, the value of $n_a^{\parallel}$ of the mean refractive index for plane polarised light vibrating parallel to the fibre axis can be obtained from the equation (Barakat and Hindeleh 1964a)

$$n_a^{\parallel} = n_L + \frac{Z^{\parallel}}{\Delta Z}\frac{\lambda}{2t_f} \tag{4.13}$$

and for $n_a^{\perp}$ (light vibrating perpendicular to the fibre axis)

$$n_a^{\perp} = n_L + \frac{Z^{\perp}}{\Delta Z}\frac{\lambda}{2t_f}. \tag{4.14}$$

For the case of a single-medium homogeneous fibre $n_s = n_c = n$ and the previous expression becomes

$$\frac{Z}{\Delta Z}\frac{\lambda}{2} = (n - n_L)t_f. \tag{4.15}$$

This formula enables us to calculate $n^{\parallel}$ and $n^{\perp}$ for a single-medium fibre.

The shape of multiple-beam Fizeau fringes across a cylindrical fibre of radius r_f immersed in a liquid of refractive index n_L, the fibre being composed of a skin of constant n_s, a core of index n_c and radius r_c, is derived as follows. Let us first derive the mathematical equation of the fringe shape in the skin region. It is to be noted that it is independent of the core characteristics (figure 4.6). Neglecting refraction and confining ourselves to $r_c \leqslant X \leqslant r_f$ we have

$$N\lambda = 2n_L t + 4Y(n_s - n_L). \tag{4.16}$$

Note that the fringe of the Nth order in the liquid region satisfies $N\lambda = 2n_L t_1$. Accordingly, if $n_s > n_L$ then the interferometric gap thickness t_1 in the liquid region is more than that in the skin region given by equation (4.16). The fringe shift in the skin region when $n_s > n_L$ is in the direction towards decreasing t. Now $Z \tan \varepsilon = t$ and

$$N\lambda - 2n_L Z \tan \varepsilon = 4(n_s - n_L)(r_f^2 - X^2)^{1/2}$$

$$-\left(Z - \frac{N\lambda}{2n_L \tan \varepsilon}\right) = \frac{4\Delta Z}{\lambda}(n_s - n_L)(r_f^2 - X^2)^{1/2}.$$

Transforming to the point $(N\lambda/2n_L \tan \varepsilon, 0)$ on the (Z, X) plane

$$-Z = \frac{4\Delta Z}{\lambda} (n_s - n_L)(r_f^2 - X^2)^{1/2}. \qquad (4.17)$$

Z here is the fringe shift measured from the point of intersection with the extension of the liquid fringe where $Z = N\Delta Z$ on the Z axis.

The shape of the fringe of Nth order over the skin region is given by

$$Z^2 \left[\left(\frac{4\Delta Z}{\lambda}\right)^2 (n_s - n_L)^2 r_f^2 \right]^{-1} + X^2/r_f^2 = 1. \qquad (4.18)$$

It is an ellipse of semi-major and semi-minor axes $(4\Delta Z/\lambda)(n_s - n_L)r_f$, and r_f on the (Z, X) plane. The direction of the fringe shift Z is determined by whether $n_s > n_L$ or $n_s < n_L$. For $n_s > n_L$ the fringe represented by half an ellipse is in the direction towards the wedge apex, while for $n_s < n_L$ the fringe is on the side of the liquid fringe away from the apex. Figures 4.6(*a*) and (*b*) show the shape of the fringes following the elliptic equation for $n_s - n_L < 0$ and $n_s - n_L > 0$, respectively. Now if $n_s - n_L$ is increased, the shape of the fringe continues to be half an ellipse on the side of the liquid fringe towards the wedge apex, but with increasing eccentricity $1 - 1/A^2$, where $A = (4\Delta Z/\lambda)(n_s - n_L)$. When $A = 1$ the fringe is a semi-circle, and if $n_s - n_L$ is decreased further the major axis becomes the minor. When $n_s = n_L$, the fringe obviously becomes a straight line in the skin region, which is the continuation of the fringe in the liquid region. By further increasing n_L, the fringe is of elliptic shape but on the other side of the liquid fringe, away from the wedge apex. This is the case for any medium of circular cross section bounded by an interface satisfying the condition of constant refractive index of the medium and its surroundings.

In the case of a cylindrical fibre composed of a core and skin immersed in a matching liquid, i.e. $n_L = n_s$, then when n_{core} is constant, the shape of the multiple-beam Fizeau fringe across the core region on the (Z, X) plane is given by

$$Z^2 \left[\left(\frac{4\Delta Z}{\lambda}\right)^2 (n_c - n_L)^2 r_c^2 \right]^{-1} + X^2/r_c^2 = 1. \qquad (4.19)$$

The semi-major and semi-minor axes are Br_c and r_c where $B = (4\Delta Z/\lambda)(n_c - n_L)$.

The effect of the magnitude of $n_s - n_L$ in the skin region has been dealt with. A similar effect of $n_c - n_L$ in the region $0 \leqslant X \leqslant r_c$, in the case of a fibre immersed in a matching liquid, takes place on the shape of the fringe.

The effect of varying the wedge angle ε on the shape of the fringe is to be seen from the values of the semi-major and semi-minor axes, namely $2(n_c - n_L)r_c/n_L \tan \varepsilon$ and r_c. As the angle ε is decreased the major axis increases, ΔZ also increases. A similar effect takes place when n_L is decreased.

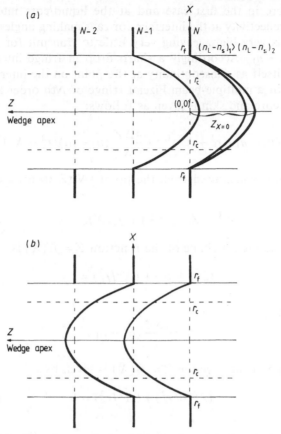

Figure 4.6 The shape of the fringes when (*a*) $(n_s - n_L) < 0$, (*b*) $(n_s - n_L) > 0$.

Consider a system of interference fringes of orders of interference N, $N + 1$ and $N + 2$ as shown in figure 4.7. The origins for the fringes

$$O_N \equiv \left(\frac{N\lambda}{2n_L \tan \varepsilon}, 0 \right)$$

$$O_{N+1} \equiv [(N + 1)\Delta Z, 0]$$

$$O_{N+2} \equiv [(N + 2)\Delta Z, 0]$$

are measured from the wedge apex. This means that the fringe systems, their shape and position provide a scanning method for detecting any defect along the fibre length at various cross sections. Changing the wedge angle ε allows scanning between any two successive orders of interference.

The fringe point at the liquid/skin or liquid/core interface in the case of matching, corresponds to the tangential ray at the circumference of the cylindrical fibre in the first case and at the liquid/core interface in the second. The reflectivity at the interface for rays making angles of incidence close to $\pi/2$ is very high, leaving very little to transmit for all values of $n_s - n_L$ or $n_c - n_L$. Accordingly a sharp drop in fringe intensity occurs, which shows itself as a discontinuity at the point on the interface.

The shape of a multiple-beam Fizeau fringe of Nth order across a fibre composed of core and skin is given as follows:

$$Z = \frac{4\Delta Z}{\lambda}(n_s - n_L)(r_f^2 - X^2)^{1/2} + \frac{4\Delta Z}{\lambda}(n_c - n_s)(r_c^2 - X^2)^{1/2}. \quad (4.20)$$

The fringe shift is measured from the point $(N\Delta Z, 0)$ towards the wedge apex

$$Z = f_1(X) + f_2(X).$$

The equation giving the shape of the function $Z = f_1(X)$ is

$$(Z^2/A^2 r_f^2) + (X^2/r_f^2) = 1 \quad (4.21)$$

where

$$A = \frac{4\Delta Z}{\lambda}(n_s - n_L)$$

while the shape of the function $Z = f_2(X)$ is given by

$$(Z^2/B^2 r_c^2) + (X^2/r_c^2) = 1 \quad (4.22)$$

where

$$B = \frac{4\Delta Z}{\lambda}(n_c - n_s).$$

The contribution of each of the two functions $f_1(X)$ and $f_2(X)$ to the formation of the fringe across the fibre is represented graphically and then addition takes place for $n_c > n_s > n_L$, for the two half ellipses on the side of the liquid fringe towards the wedge apex.

For $n_c > n_s$ and $n_L > n_s$, $Z = f_2(X) - f_1(X)$. Figures 4.7(a) and (b) show the shape of the fringe across the core and skin regions for the two cases. From figure 4.7(a) the fringe shift Z at $X = 0$ is given by

$$Z_{X=0} = \frac{4\Delta Z}{\lambda}[(n_s - n_L)r_f + (n_c - n_s)r_c] \quad (4.23)$$

while from figure 4.7(b)

$$Z_{X=0} = \frac{4\Delta Z}{\lambda}[(n_c - n_s)r_c - (n_L - n_s)r_f]. \quad (4.24)$$

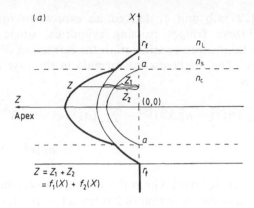

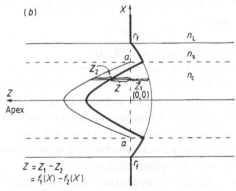

Figure 4.7 The shape of the fringes across the core and skin where (*a*) $n_s > n_L$ and $n_c > n_s$, (*b*) $n_L > n_s$ and $n_c > n_s$.

4.2.2 Multilayer fibres with regular transverse sections

The mathematical expression for the shape of a multiple-beam Fizeau fringe crossing a cylindrical fibre having multilayers is obtained by adding the appropriate terms to the expression for the optical path length (OPL) given by equation (4.6). Such terms represent the contributions of each layer constituting the multilayer fibre. This treatment leads to an expression as follows:

$$\left(\frac{Z}{\Delta Z}\right)_x \frac{\lambda}{2} = 2\left(\sum_{k=1}^{m} n_k r_k - \sum_{k=1}^{m-1} n_k r_{k+1} - n_L r_1\right) \qquad (4.25)$$

where n_k denotes the refractive index of the layer k, $k = 1, 2, \dots m$, r_k is its radius, n_L is the refractive index of the enclosed liquid and r_1 the radius of the outer layer of the fibre. This is the equation for multilayer fibres, at $X = 0$, reported by El-Nicklawy and Fouda (1980a) and Hamza and Kabeel (1986).

El-Hennawi (1988a,b and c) derived an expression for the shape of multiple-beam Fizeau fringes crossing cylindrical single-, double- and multilayer fibres taking into consideration the refraction of the optical ray through the fibre. The mathematical formula in the case of a multilayer cylindrical fibre is

$$\left(\frac{Z}{\Delta Z}\right)_x \frac{\lambda}{2} = 2\left(\sum_{k=1}^{m} (n_k^2 r_k^2 - n_L^2 X^2)^{1/2} - \sum_{k=1}^{m-1} (n_k^2 r_{k+1}^2 - n_L^2 X^2)^{1/2}\right.$$

$$\left. - n_L(r_1^2 - X^2)^{1/2}\right) \quad (4.26)$$

m being the number of layers within the fibre cross section. At $X = 0$, no refraction takes place and equation (4.26) leads to equation (4.25).

4.2.3   Multiple-beam Fizeau fringes applied to fibres with irregular transverse sections

Homogeneous fibres (Hamza et al 1985a)
Simmens (1958) described a technique using the Babinet compensator to determine the birefringence in objects of constant weight per unit length but irregular cross sectional shape. To determine the refractive indices and birefringence of fibres with irregular transverse sections Hamza (1980) described a method using two-beam interference and scanning electron microscopy. In the following, multiple-beam Fizeau fringes are applied to homogeneous fibres with irregular transverse sections.

Figure 4.8 shows a fibre with irregular transverse section introduced in a silvered liquid wedge. The transverse sectional area of the fibre in the (X, Y) plane is A, given by

$$A = \int_K^L (Y_1 - Y_2)\, dX \quad (4.27)$$

where Y_1 and Y_2 are the intersection points of a scanning line parallel to the Y axis, with the circumference of the fibre cross section. The line lies between C and D, from $X = K$ to $X = L$.

The optical path length (OPL) of the ray BB′ is given by

$$\text{OPL} = [t - (Y_1 - Y_2)]n_L + (Y_1 - Y_2)n_a \quad (4.28)$$

and on a fringe of interference order N

$$N\lambda = 2n_L t + 2(n_a - n_L)(Y_1 - Y_2). \quad (4.29)$$

Substituting for $t = Z \tan \varepsilon$ and transforming the origin to the point $(N\lambda/2n_L \tan \varepsilon, 0)$ on the (Z, X) plane we have

$$n_L \tan \varepsilon\, Z = (n_a - n_L)(Y_1 - Y_2) \quad (4.30)$$

where Z is the new value after transforming the origin. It measures the fringe shift of the Nth order in the fibre region from its position in the liquid region.

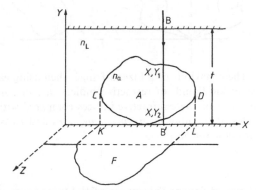

Figure 4.8 A fibre with irregular transverse section immersed in a silvered liquid wedge. The cross sectional area of the fibre is A and the refractive index of the liquid is n_L. (From Hamza *et al* 1985a.)

Integrating over the region $L \geqslant X \geqslant K$ gives the area under the fringe shift F

$$\int_K^L (Y_1 - Y_2)\, dX = \frac{n_L \tan \varepsilon}{n_a - n_L} \int_K^L Z\, dX. \qquad (4.31)$$

Let

$$\int_K^L Z\, dX = F \qquad \text{and} \qquad A = \frac{n_L \tan \varepsilon}{n_a - n_L} F.$$

Therefore

$$n_a - n_L = \frac{F}{2A} \frac{\lambda}{\Delta Z} \qquad (4.32)$$

and for polarised light vibrating parallel to the fibre axis

$$n_a^{\parallel} = n_L + \frac{F^{\parallel}}{2A} \frac{\lambda}{\Delta Z} \qquad (4.33)$$

with a similar formula for $n_a^{\perp}$. The mean refractive indices $n_a^{\parallel}$ and $n_a^{\perp}$ of the fibre can be readily calculated after determining the values of F and A. This also applies to the birefringence Δn_a given by

$$\Delta n_a = \frac{F^{\parallel} - F^{\perp}}{A} \frac{\lambda}{2\Delta Z}. \qquad (4.34)$$

The behaviour of the Fizeau fringe when using immersion liquids of different refractive indices is shown in figure 4.9.

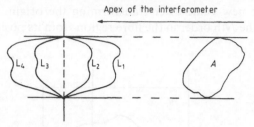

Apex of the interferometer

Figure 4.9 The behaviour of a Fizeau fringe when using each of four different immersion liquids of refractive indices n_{L_1}, n_{L_2}, n_{L_3} and n_{L_4}. The relative values of these refractive indices compared with the mean refractive index of the fibre n_a are $n_{L_1} > n_{L_2} > n_a$ and $n_a > n_{L_3} > n_{L_4}$. (From Hamza *et al* 1985a.)

To eliminate the need for measurement of the transverse sectional area of the fibre, two different immersion liquids are used, having refractive indices n_{L_1} and n_{L_2} at the same temperature (Hamza *et al* 1986). The following two equations enable determination of the mean refractive index n_a

$$\frac{F_1}{2A} \frac{\lambda}{\Delta Z_1} = n_a - n_{L_1} \qquad (4.35)$$

and

$$\frac{F_2}{2A} \frac{\lambda}{\Delta Z_2} = n_a - n_{L_2} \qquad (4.36)$$

where F_1 and F_2 are the areas enclosed under the fringe shift and ΔZ_1 and ΔZ_2 are the interfringe spacings for the two immersion liquids.

Heterogeneous fibres with skin/core structure (Hamza et al *1985b)*
Figure 4.10 shows a cross section of a fibre having an irregular cross sectional core surrounded by an irregular skin. The transverse sectional area of the fibre in the (X, Y) plane is A given by

$$A = \int_M^S (Y_1 - Y_2)\, dX \qquad (4.37)$$

where Y_1 and Y_2 are the intersection points of a line parallel to the Y axis, with the circumference of the fibre cross section, $S \geqslant X \geqslant M$. The transverse sectional area of the core in the (X, Y) plane is B given by

$$B = \int_P^Q (Y_3 - Y_4)\, dX$$

where Y_3 and Y_4 are the intersection points of the scanning line parallel to the Y axis, with the circumference of the core cross section, $Q \geqslant X \geqslant P$.

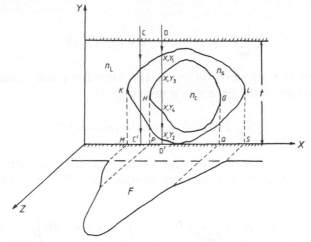

Figure 4.10 A fibre with irregular transverse section having a core surrounded by a skin immersed in a silvered liquid wedge. The area enclosed under the fringe shift is F. (From Hamza *et al* 1985b.)

The optical path length (OPL) of the ray DD′ is given by

$$\text{OPL} = [t - (Y_1 - Y_2)]n_L + [(Y_1 - Y_2) - (Y_3 - Y_4)]n_s + (Y_3 - Y_4)n_c \tag{4.38}$$

and on a fringe of interference order N

$$N\lambda = 2n_L t + 2(n_s - n_L)(Y_1 - Y_2) + 2(Y_3 - Y_4)(n_c - n_s). \tag{4.39}$$

Transforming the origin to the point $(N\lambda/2n_L \tan \varepsilon, 0)$ on the (Z, X) plane we have

$$n_L \tan \varepsilon \, Z = (n_s - n_L)(Y_1 - Y_2) + (n_c - n_s)(Y_3 - Y_4).$$

Integrating the above equation in the region $M \geqslant X \geqslant S$ gives the area under the fringe shift F

$$n_L \tan \varepsilon \int_M^S Z \, dX = (n_s - n_L) \int_M^S (Y_1 - Y_2) \, dX$$

$$+ (n_c - n_s) \int_P^Q (Y_3 - Y_4) \, dX \tag{4.40}$$

where

$$\int_M^S Z \, dX = F$$

$$Fn_L \tan \varepsilon = (n_s - n_L)A + (n_c - n_s)B$$

and

$$\frac{\lambda}{2\Delta Z} F = (n_\mathrm{s} - n_\mathrm{L})A + (n_\mathrm{c} - n_\mathrm{s})B.$$

The value of the refractive index of the fibre for plane polarised light vibrating parallel to the fibre axis is given by

$$\frac{\lambda}{2\Delta Z} F^{\parallel} = (n_\mathrm{s}^{\parallel} - n_\mathrm{L})A + (n_\mathrm{c}^{\parallel} - n_\mathrm{s}^{\parallel})B \qquad (4.41)$$

with a similar formula for the perpendicular direction of light vibration, leading to an expression for the birefringence $\Delta n_\mathrm{c} = n_\mathrm{c}^{\parallel} - n_\mathrm{c}^{\perp}$ of the core as follows:

$$\Delta n_\mathrm{c} = B^{-1}\left(\frac{\lambda}{2\Delta Z}(F^{\parallel} - F^{\perp}) - \Delta n_\mathrm{s}(A - B)\right). \qquad (4.42)$$

A mean refractive index of the fibre, n_a, can be deduced by putting $n_\mathrm{s} = n_\mathrm{c} = n_\mathrm{a}$.

Multilayer fibres (Hamza and Kabeel 1986)

Figure 4.11 shows a cross section of a cylindrical multilayer fibre, introduced in a silvered liquid wedge. The fibre has m layers of refractive indices $n_1, n_2, ..., n_m$, where n_1 is the refractive index of the outer layer and $n_m = n_\mathrm{c}$. The radii of the fibre layers are represented by $r_Q = (X^2 + Y^2)^{1/2}$, $Q = 1, 2, ..., m$. The optical path length (OPL) of the EE' is given by

$$\mathrm{OPL} = (t - 2Y_1)n_\mathrm{L} + 2(Y_1 - Y_2)n_1$$
$$+ 2(Y_2 - Y_3)n_2 + \cdots 2(Y_{m-1} - Y_m)n_{m-1} + 2Y_m n_m. \qquad (4.43)$$

From the basic relation of interference

$$N\lambda - 2n_\mathrm{L}Z \tan \varepsilon = 4 \sum_{Q=1}^{m} (n_Q - n_{Q-1})Y_Q. \qquad (4.44)$$

Transforming the origin to the point $(N\lambda/2n_\mathrm{L} \tan \varepsilon, 0)$ on the (Z, X) plane

$$n_\mathrm{L} \tan \varepsilon\, Z = 2 \sum_{Q=1}^{Q=m} (n_Q - n_{Q-1})Y_Q.$$

Z is now the new value and its direction is towards the wedge apex. The shape of the Fizeau fringes on the (Z, X) plane is given by

$$\frac{\lambda}{4\Delta Z} Z = \sum_{Q=1}^{m} (n_Q - n_{Q-1})(r_Q^2 - X^2)^{1/2}. \qquad (4.45)$$

The X value defines the number of layers. To overcome the difficulty caused by the presence of any irregularity in the transverse sectional shape, the area

enclosed by the deviation of a fringe as it crosses the fibre perpendicular to
its axis is considered to represent the path difference integrated across this
fibre (see Simmens 1958). The part of the area enclosed under the fringe
deviation F_m in the (Z, X) plane corresponding to a multilayer fibre in a
region defined by $\alpha \leqslant X \leqslant \beta$, is related to the parts of cross sectional areas
of the fibre layers $A_{Q,m}$ by the equation

$$\frac{\lambda}{4\Delta Z} F_m = \sum_{Q=1}^{m} (n_Q - n_{Q-1})A_{Q,m} \tag{4.46}$$

where

$$F_m = \int_{\alpha}^{\beta} Z \, \mathrm{d}X$$

and

$$A_{Q,m} = \int_{\alpha}^{\beta} (r_Q^2 - X^2)^{1/2} \, \mathrm{d}X$$

and α and β are two chosen points on the X axis of the interferogram.

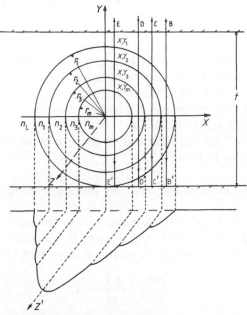

Figure 4.11   A cross section of a cylindrical multilayer fibre introduced
in a silvered liquid wedge. The shape of the interference fringe shift
inside every layer of the fibre is shown in the profile given at the base of
the figure. (From Hamza and Kabeel 1986.)

4.3   Optical fibres: step-index and graded-index

4.3.1   Multiple-beam interference fringes applied to step-index optical fibres to determine fibre characteristics

Step-index and graded-index optical fibres are used as optical waveguides in optical communication systems. They are cylindrical fibres of radius r_f composed of a skin, usually termed clad, of refractive index n_{clad} and a core of refractive index n_{core} and radius $r_c = a$, $n_{core} > n_{clad}$.

For step-index optical fibres, n_{clad} and n_{core} are of constant values for the fibre. They are either monomode or multimode, the difference being in the dimensions of core and clad. For a monomode step-index, $2r_c \approx 10\ \mu m$ or less, while $2r_f = 125\ \mu m$. A multimode step-index fibre has $2r_c \approx 80\ \mu m$ and $2r_f = 125\ \mu m$. In all types of optical fibres used as waveguides, the value of n_{clad} remains constant, while n_{core} is constant only for step-index fibres. For graded-index fibres, n_{core} varies with the distance from the fibre centre according to a power law.

For a step-index fibre, the equation giving the fringe shift Z measured from the point $(N\Delta Z, 0)$ towards the wedge apex, when the optical fibre is introduced in a silvered liquid wedge is as derived before (equation (4.20))

$$Z = \frac{4\Delta Z}{\lambda}(n_{clad} - n_L)(r_f^2 - X^2)^{1/2} + \frac{4\Delta Z}{\lambda}(n_{core} - n_{clad})(r_{clad}^2 - X^2)^{1/2}$$

$$= f_1(X) + f_2(X).$$

The shape of the function $Z = f_1(X)$ is given by equation (4.21) describing an ellipse of semi-major and semi-minor axes Ar_f and r_f where $A = 4\Delta Z(n_{clad} - n_L)/\lambda$, while the elliptic shape of the function $f_2(X)$ is of semi-major and semi-minor axes Br_c and r_c where $B = 4\Delta Z(n_{core} - n_{clad})/\lambda$ in accordance with equation (4.22).

The addition of the contributions of the two functions is as explained before and appears in figure 4.7.

Deduction of the index profile of a step-index optical fibre from the fringe shift
As derived, the shape of multiple-beam Fizeau fringes across a cylindrical fibre in the clad region $r_c \leqslant X \leqslant r_f$ is given by equation (4.21). Therefore the equation of the tangent to the ellipse at any point (Z', X') is

$$\frac{ZZ'}{A^2 r_f^2} + \frac{XX'}{r_f^2} = 1.$$

The slope of the straight line is

$$\frac{dX}{dZ} = -\frac{Z'}{X'}\frac{1}{A^2}.$$

Therefore

$$\frac{dX}{dZ}\frac{X'}{Z'} = -[\lambda^2/16(\Delta Z)^2(n_{clad} - n_L)^2]$$

$$= \text{constant for the fringe system.}$$

This is the parameter which indicates the constancy of the refractive index characterising a step-index fibre. Rearranging

$$n_{clad} - n_L = \lambda/4\Delta Z \left| \frac{dX}{dZ}\frac{X'}{Z'} \right|^{1/2} \qquad r_c \leqslant X \leqslant r_f.$$

For the core region, when a matching liquid is used in the silvered liquid wedge, $n_L = n_{clad}$, then

$$n_{core} - n_L = \lambda/4\Delta Z \left| \frac{dX}{dZ}\frac{X'}{Z'} \right|^{1/2} \qquad 0 \leqslant X \leqslant r_c.$$

Figure 4.12 shows the index profile $n_{clad} - n_L$ over $r_c \leqslant X \leqslant r_f$ and $n_{core} - n_L$ over $0 \leqslant X \leqslant r_c$, in the case of a matching liquid.

For the general case of a step-index fibre inserted in a silvered liquid wedge, $n_s \neq n_L$, the index profile over the core region can be deduced from experimentally recorded fringes by subtracting the contribution of the clad in the core region from the value of Z on the fringe for all values of X, for the case $n_c \geqslant n_{clad} \geqslant n_L$. For the second case, when $n_L > n_{clad}$, addition of $|Z_1|$ to Z from the experimental curve gives Z_2 for all X. The contribution of the clad is obtained by merely extending the elliptic fringe in the clad region to the core region. Then by determining $|(dX/dZ)(X'/Z')|^{1/2}$ over the core region $-a \leqslant X \leqslant a$, $a = r_c$, $(n_{core} - n_{clad})$ for all X, the index profile can be determined.

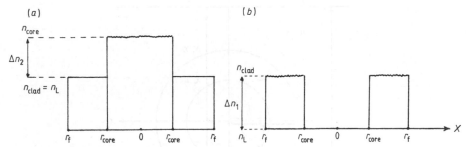

Figure 4.12 The index profile: (*a*) the case of a matching liquid and (*b*) n_{clad} over the clad region.

For the clad, Z values in the clad region $r_c \leqslant X \leqslant r_f$ on both sides of $Z = 0$ give $n_{clad} - n_L$ from

$$\left| \frac{dX}{dZ}\frac{X'}{Z'} \right|^{1/2} \frac{\lambda}{4\Delta Z}.$$

4.3.2   The theory of multiple-beam Fizeau fringes applied to graded-index optical fibres

Marhic *et al* (1975) applied two-beam interference to optical fibres immersed in an index matching oil and illuminated perpendicular to the fibre axis. They showed that simple analytical expressions for the optical path difference are obtained for fibres whose index profile is a quadratic function of the radius of the core. Saunders and Gardner (1977) applied the Marhic method to the more general class of graded-index optical fibre. In this case Δn was calculated from the maximum fringe shift, and using a computer program α was calculated from any point on the fringe.

The following gives the mathematical equation of the family of Fizeau fringes across a graded-index optical waveguide (Barakat *et al* 1985). The fibre is assumed to be of a perfectly circular cross section and to have a core surrounded by a cladding. The fibre is introduced in a silvered liquid wedge. The fibre axis is adjusted perpendicular to the edge of the wedge. Let us assume that one of the two silvered optical flats forming the wedge, the lower one, just touches the circumference of the fibre. Figure 4.13 represents a cross section of a cylindrical graded-index fibre of radius r_f having a core of variable refractive index $n(r)$ and radius r_c and a cladding of constant refractive index n_{clad}

$$n(r) = n(0)[1 - 2\Delta(r/a)^\alpha]^{1/2} \qquad 0 \leqslant r \leqslant a \qquad (4.47)$$

where r is the distance from the centre of the core and the core radius is a (Gloge and Marcatili 1973), $\Delta = (n^2(0) - n^2(a))/(2n^2(0))$ and α is a para-

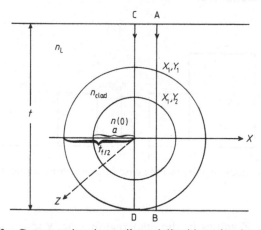

Figure 4.13   Cross section in a silvered liquid wedge interferometer with a graded-index waveguide fibre of variable index core $n(r)$. (From Barakat *et al* 1985.)

meter that determines the shape of the profile. The cylindrical graded-index fibre is immersed in a liquid enclosed between two optical flats ($\pm \lambda/50$) with highly reflecting partially transparent silver layers. They form a silvered liquid wedge; the liquid refractive index n_L is close to n_{clad}. A parallel beam of monochromatic light of wavelength λ represented by AB and CD is incident normal to the lower component of the wedge, the wedge angle ε being small to fulfil the necessary conditions of multiple-beam fringe formation. The fibre axis is chosen as the z axis, and the edge is parallel to the x axis. The thickness of the interferometric gap is represented by t.

For the optical path length (OPL) of the ray AB

$$\text{OPL} = (t - 2y_2)n_L + 2(y_2 - y_1)n_{clad}$$

$$+ 2 \int_0^{y_1 = (a^2 - x_1^2)^{1/2}} n(r)\, \mathrm{d}y \tag{4.48}$$

where $n(r)$ is defined by equation (4.47). For $\lambda \ll 1$, which is the case for graded-index fibres, $n(r) = n(0) - \Delta n(r/a)^\alpha$, where $\Delta n = (n(0) - n(a))$. Therefore

$$\text{OPL} = (t - 2y_2)n_L + 2(y_2 - y_1)n_{clad} + 2n(0)(a^2 - x_1^2)^{1/2}$$

$$- 2\frac{\Delta n}{a^\alpha} \int_0^{(a^2 - x_1^2)^{1/2}} (x_1^2 + y^2)^{\alpha/2}\, \mathrm{d}y. \tag{4.49}$$

On a fringe of order of interference N

$$N\lambda = 2(\text{OPL}) = 2n_L t + 4y_2(n_{clad} - n_L) + 4\Delta n y_1$$

$$- \frac{4\Delta n}{a^\alpha} \int_0^{(a^2 - x_1^2)^{1/2}} (x_1^2 + y^2)^{\alpha/2}\, \mathrm{d}y. \tag{4.50}$$

Now $t = z \tan \varepsilon$ ($z = 0$ at $t = 0$) so

$$N\lambda - 2n_L z \tan \varepsilon = 4y_2(n_{clad} - n_L) + 4\Delta n y_1$$

$$- \frac{4\Delta n}{a^\alpha} \int_0^{(a^2 - x_1^2)^{1/2}} (x_1^2 + y^2)^{\alpha/2}\, \mathrm{d}y. \tag{4.51}$$

Transforming to the point $(0, N\lambda/2n_L \tan \varepsilon)$, we have

$$z2n_L \tan \varepsilon = 4y_2(n_{clad} - n_L) + 4\Delta n y_1 - 4\frac{\Delta n}{a^\alpha} \int_0^{(a^2 - x_1^2)^{1/2}} (x_1^2 + y^2)^{\alpha/2}\, \mathrm{d}y.$$

$$\tag{4.52}$$

Now Δz is the spacing between any two consecutive fringes in the liquid region and is equal to $\lambda/2n_L \tan \varepsilon$. If δz denotes the fringe shift of the Nth order in the fibre region from its position in the liquid region, we have

$$\left(\frac{\delta z}{\Delta z}\right)_{x_1} \frac{\lambda}{2} = 2\left(y_2(n_{\text{clad}} - n_{\text{L}}) + \Delta n y_1 - \frac{\Delta n}{a^\alpha} \int_0^{(a^2 - x_1^2)^{1/2}} (x_1^2 + y^2)^{\alpha/2} \, dy\right)$$

$$= 2\left((n_{\text{clad}} - n_{\text{L}})\sqrt{r_f^2 - x_1^2} + \Delta n\sqrt{a^2 - x_1^2}\right.$$

$$\left. - \frac{\Delta n}{a^\alpha} \int_0^{(a^2 - x_1^2)^{1/2}} (x_1^2 + y^2)^{\alpha/2} \, dy\right). \tag{4.53}$$

This gives the required equation giving $(\delta z/\Delta z)$ for any value of x_1 where $0 \leqslant x_1 \leqslant a$ in terms of Δn and α. Substituting $x_1 = 0$ gives the following expression:

$$\left(\frac{\delta z}{\Delta z}\right)\frac{\lambda}{2} = (n_{\text{clad}} - n_{\text{L}})t_f + t_{\text{core}}\Delta n \frac{\alpha}{(\alpha + 1)} \tag{4.54}$$

where $t_{\text{core}} = 2a$ and $t_f = 2y_2$. A similar form of equation (4.54) was reported by Saunders and Gardner (1977) for two-beam interference when $n_{\text{clad}} = n_{\text{L}}$.

In contrast to the case of step-index fibres when $\alpha = \infty$, we have reached the following equation (Barakat 1971):

$$\left(\frac{\delta z}{\Delta z}\right)\frac{\lambda}{2} = (n_{\text{clad}} - n_{\text{L}})t_f + t_{\text{core}}(n_{\text{core}} - n_{\text{clad}}). \tag{4.55}$$

Substituting two values of x_1 in equation (4.53) an expression is derived giving the parameter α in terms of x_1, x_2, $(\delta z/\Delta z)_{x_1}$, $(\delta z/\Delta z)_{x_2}$, n_{clad} and n_{L}, which allows us to calculate α numerically. Then substituting in equation (4.53) any value of x where $0 \leqslant x \leqslant a$ gives the value of Δn.

A more general approach to calculating both α and Δn making use of more than two values of $\delta z/\Delta z$ has been adopted. The procedure is a minimum variance technique in which

$$I_\alpha(x) = \int_0^{(a^2 - x^2)^{1/2}} (x^2 + y^2)^{\alpha/2} \, dy$$

is evaluated numerically.

In equation (4.53), Δn and α are initially unknown quantities. The object is to fit the experimentally measured fringe profile to the above equation to determine both Δn and α. For this purpose, it is necessary to normalise the theoretical profile to some experimental points to account for the response of the measuring device (magnification and other factors). The point selected is that for which $F(x) = \delta z$ is independent of Δn and α.

Let

$$F(a) = \frac{4\Delta z}{\lambda} \left[(n_{\text{clad}} - n_{\text{L}})(r_f^2 - a^2)^{1/2}\right]$$

be the value of the profile function at $x = a$. The normalised profile function may be written as

$$F^*(x) = \frac{F(x)}{F(a)} \delta z_{\text{expt}} \Big|_{x=a}.$$

By varying α and scanning through Δn for each value of α, one can locate the particular values of α and Δn giving best fit to the experimental profile.

4.4 The theory of fringes of equal chromatic order (FECO) applied to fibres

Multiple-beam Fizeau fringes formed when a parallel beam of mono-chromatic light is incident on a silvered wedge are known to be localised on a plane close to the wedge which is the Feussner surface of zero order and with a position independent of wavelength. Now, if the Feussner surface is projected on the slit of a spectrograph and a white light source replaces the monochromatic beam, a family of white light fringes of equal chromatic order (Tolansky 1960) is seen on the spectral plane. The shape of the resulting fringes depends basically on the way in which $n_\lambda t$ varies along the section of the interferometer selected by the slit, t being the thickness of the interferometric gap and n_λ being the refractive index of the enclosed medium. If the line is taken to be the x axis, then, in general, $n_\lambda t = f(x)$. The spectral plane is the (λ, x) plane and, thus, the shape of the fringes results from transforming the relation $n_\lambda t = f(x)$ from the (t, x) plane to the (λ, x) plane by using the relation $N\lambda = 2n_\lambda t$ for bright fringes in transmission and dark fringes in reflection, neglecting phase changes in reflection. For a silvered liquid wedge, adjusted with its edge parallel to the slit, $n_{L,\lambda} t = K$ (a constant for a certain λ), $n_{L,\lambda}$ being the refractive index of the enclosed liquid. The family of fringes of equal chromatic order on the (λ, x) plane is represented by $\lambda_N = 2K/N$, where the order of interference $N = 1, 2, 3, \ldots$. Accordingly, the fringes are straight lines vertical and parallel to the x axis, with the wavenumber separation $\Delta \nu = \nu_{(N+1)} - \nu_N = \frac{1}{2} n_{L,\lambda'} t$, where $n_{L,\lambda'}$ is the average refractive index of the enclosed liquid for the wavelength range $\lambda_{(N+1)} - \lambda_N$.

Let us consider the region of the interferometer enclosing the fibre. The OPL of the central beam has been calculated and the condition for a bright fringe is given by equation (4.7). On the (λ, x) plane, t is a constant, as the edge of the wedge is adjusted parallel to the slit. Squaring equation (4.7) and transforming the origin to $(2n_{L,\lambda} t/N, 0)$, we have

$$\lambda^2 = (16/N^2)[(n_{s,\lambda} - n_{L,\lambda})^2(r_f^2 - x^2) + (n_{c,\lambda} - n_{s,\lambda})^2(r_c^2 - x^2)$$
$$+ 2(n_{s,\lambda} - n_{L,\lambda})(n_{c,\lambda} - n_{s,\lambda})(r_f^2 - x^2)^{1/2}(r_c^2 - x^2)^{1/2}]. \quad (4.56)$$

This is the required equation for the family of FECO fringes crossing the fibre in the region $0 \leqslant x \leqslant r_c$ and $-r_c \leqslant -x \leqslant 0$, where $n_{s,\lambda}$, $n_{c,\lambda}$ and $n_{L,\lambda}$

are the refractive indices of the skin, the core of fibre and of the enclosed liquid, respectively, at the wavelength corresponding to any point on the fringe of order N.

For $x = 0$

$$\lambda = (4/N)(n_{s,\lambda} - n_{L,\lambda})r_f + (n_{c,\lambda} - n_{s,\lambda})r_c$$
$$= (2/N)(n_{s,\lambda}t_s + n_{c,\lambda}t_c - n_{L,\lambda}t_f)$$
$$= (2/N)(n_{a,\lambda} - n_{L,\lambda})t_f$$

where $n_{a,\lambda}$ is the average refractive index, defined in equation (4.11), for the wavelength of the point on the fringe of order N when $x = 0$. For example, λ at point b on the fringe of order $N + 1$ (figure 4.14) shows a graphical representation of equation (4.56).

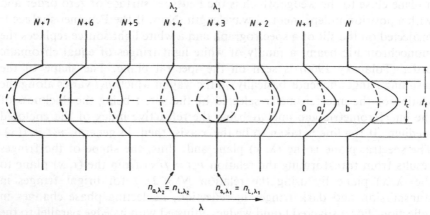

Figure 4.14 Graphical presentation of FECO across a fibre immersed in a silvered liquid wedge. The broken curves correspond to a single-medium fibre, the full curves correspond to a fibre of thickness t_f having a core of thickness t_c surrounded by a skin. For the fringe of order $N + 3$, $n_{s,\lambda_1} = n_{L,\lambda_1}$ as is shown, while for the fringe of order $N + 4$, $n_{a,\lambda_2} = n_{L,\lambda_2}$. (From Barakat 1971.)

The point $(2n_{L,\lambda}t/N, 0)$ is the point of intersection of the extension of the straight-line portion of the fringe of order N in the liquid region with the fibre axis. It is represented by the point 0 in figure 4.14 for the fringe of order $N + 1$.

For the case of a fibre having a single medium, $n_{s,\lambda} = n_{c,\lambda} = n_\lambda$, the shape of the fringes is obtained by substituting in equation (4.56). It is identical to that for the fringes crossing the fibre in the region $r_c \leqslant x \leqslant r_f$. The fringes are represented by

$$\lambda^2/(4/N)[(n_\lambda - n_{L,\lambda})r_f]^2 + x^2/r_f^2 = 1.$$

This represents a family of ellipses of semi-major and semi-minor axes $4(n_\lambda - n_{L,\lambda})r_f/N$ and r_f, for successive integer values of N. For a fibre of a single medium and of refractive index n_λ, the length of the principal axis of any ellipse along the λ axis varies with $(n_\lambda - n_{L,\lambda})/N$, where n_λ and $n_{L,\lambda}$ are the refractive indices of the fibre and the enclosed liquid, respectively, for the wavelength determined by the apex of the ellipse, i.e. on the fringe of order N for $x = 0$. In figure 4.14 this wavelength corresponds to the point a on the broken curve of interference order $N + 1$. The three cases to be considered are shown in figure 4.14 by the broken curves:

(*a*) $n_\lambda > n_{L,\lambda}$. This is the case for the fringes of order N, $N + 1$ and $N + 2$.

(b) $n_{\lambda_1} = n_{L,\lambda_1}$. No fringe shift occurs between the fringe in the liquid region and as it crosses the fibre in the region $r_c \leqslant x \leqslant r_f$. In figure 4.14 this occurs at the fringe of interference order $N + 3$, which determines λ_1.

(c) $n_\lambda < n_{L,\lambda}$. The fringes are convex towards shorter wavelengths. This is the case for the fringes of order $N + 4$, $N + 5$ and $N + 6$.

In the presence of skin and core for $0 \leqslant x \leqslant r_c$ and its mirror image in the fibre axis, the fringes are represented by the full curves in figure 4.14. When $n_{L,\lambda_1} = n_{s,\lambda_1}$, the fringe shift at $x = 0$, between the liquid fringe and the fringe across the fibre equals $(n_{c,\lambda_1} - n_{s,\lambda_1})4r_c/N$. Now, if $n_{c,\lambda} > n_{s,\lambda}$, then the appearance of the fringe of equal chromatic order in this region is as represented by the full curve corresponding to the fringe of order $N + 3$ in figure 4.14. If $n_{a,\lambda} = n_{L,\lambda}$, where $n_{a,\lambda} = (n_{c,\lambda}t_c + n_{s,\lambda}t_s)/t_f$, no fringe shift takes place at $x = 0$ between the liquid fringe and that across the fibre. This is represented by the full curve of the fringe of order $N + 4$ in figure 4.14.

Faust (1954) applied FECO to the determination of the mean refractive indices of fibres. His experimental approach was based on the use of points of zero shift on fringes.

4.5   Applications of multiple-beam interferometric methods to the determination of some physical properties of fibres

The refractive indices of synthetic and natural fibres for plane polarised light vibrating parallel and perpendicular to the fibre axis provide a convenient measure of the extent of molecular alignment with respect to the fibre axis. Measurement of the birefringence of the skin and core of the fibre provides a measure of the degree of scatter of molecules about preferred orientations. The degree of molecular orientation and the closeness of molecular packing at different regions of anisotropic fibres assist in investigating the fibre structure. Multiple-beam interferometry is a very useful tool in fibre science. For synthetic and natural fibres interferometry permits accurate measurement of refractive indices $n^\parallel$ and $n^\perp$ for the fibre core and skin and the fibre birefringence. Multiple-beam interferometric measurement of the variation of refractive indices of fibres with (i) wavelength of light, $dn/d\lambda$ (dispersion), (ii) temperature, dn/dT (opto-thermal) and (iii) tensile properties, e.g. elongation per unit length (opto-mechanical), can also be achieved. Multiple-beam fringes also provide quantitative information on the optical properties of the skin and core of heterogeneous fibres and the variation of these properties along the fibre axis. Interferometric applications deal with natural and synthetic fibres with regular and irregular transverse sections and twisted fibres.

For optical fibres, multiple-beam interferometry allows determination of the index profile across the clad and core of step-index and graded-index fibres. It also provides information about the fibre characteristics related to their structure, the multilayer structure of the fibre core and the parameter α, which governs the variation of the refractive index of core with distance from fibre centre. Such information assists in controlling the process of fibre manufacture by modified chemical vapour deposition (MCVD).

The interference fringes produced from multiple-beam interferometric methods are extremely sharp and the accuracy in measuring the fringe shift is remarkably high. The fringe displacement is proportional to twice the phase difference introduced by the fibre. Therefore multiple-beam interferometric methods are much more sensitive than methods using two-beam interference fringes (Tolansky 1948).

To complete the picture of the molecular arrangement within fibres, these optical studies are often made in conjunction with investigations by means of x-ray diffraction, electron microscopy and molecular absorption spectroscopy.

The investigation of the surface topography of different fibres using interferometric techniques is dealt with in Chapter 5.

4.5.1   *Experimental arrangement and procedure for forming multiple-beam interference fringes*

Using a wedge interferometer produced by Tolansky (1948, 1960) multiple-beam localised Fizeau fringes were used in the determination of the mean refractive indices and birefringence of fibres. Measurements of optical path length differences are performed by introducing a sample of the fibre in a silvered liquid wedge.

Faust (1952, 1954) described a method for the application of multiple-beam interference microscopy to the determination of refractive index variations within optically heterogeneous specimens. He determined the mean refractive index of the fibre using white light fringes of equal chromatic order.

Following Barakat and El-Hennawi (1971), a parallel beam of plane polarised monochromatic light is used to illuminate a wedge interferometer placed on a microscope stage, the incidence being normal. Figure 4.15 shows the optical system used to produce multiple-beam Fizeau fringes in transmission and at reflection, respectively. The wedge interferometer consists of two circular optical flats, 35 mm in diameter and 7 mm thick and flat to ± 0.01 μm. To produce multiple-beam Fizeau fringes in transmission, the inner surface of each flat is coated with a highly reflecting and partially transmitting silver layer. This is prepared by thermal evaporation at low pressure of less than 10^{-5} Torr. Its reflectance is more than 75% with transmittance of about 22%. In the case of the reflecting system, the reflectance of the lower optical flat is more than 90% and that of the upper flat is about 70%. The two coated optical flats are then put in a holder and a drop of an immersion liquid having a refractive index close to that of the fibre (measured by the immersion method) is put on the silvered face of the lower optical flat. The fibre is then immersed in the liquid and its ends fixed. The upper flat is then introduced to form the silvered liquid wedge. In interferometric investigations other than dispersion, it is recommended that multilayer coatings of low (L) and high (H) refractive index dielectric layers LHLH...L are used instead of silver to enhance reflectivity without increasing absorption. This leads to sharper fringes of much smaller half width. On the microscope stage, both the gap thickness and wedge angle are adjusted by using three screws to form the sharpest fringes across the fibre. The phase lag that exists between successive reflecting beams, which is a

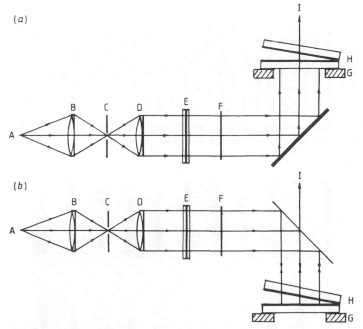

Figure 4.15 Optical set-up for producing multiple-beam Fizeau fringes in transmission (a) and at reflection (b).

A, mercury lamp; B, condenser lens; C, iris diaphragm; D, collimating lens; E, polariser; F, monochromatic filter; G, microscope stage; H, silvered liquid wedge interferometer; I, to camera attached to the microscope.

deviation from a constant value that increases with the order of the reflected beam, has to be suppressed. The phase lag $\delta = (2\pi/\lambda)\frac{4}{3} n_L t \varepsilon^2$. When this is achieved it allows a larger number of effective multiple-reflected beams to contribute to fringe formation. Thus sharper fringes with smaller half width as fraction of order separation are secured. For synthetic and natural fibres $t_f < 100$ μm, which facilitates the selection of small values of the interferometric gap t and small wedge angle ε, using circular flats of 35 mm in diameter. But in optical fibres, the fibre thickness for step-index and graded-index fibres is ≈ 125 μm, and wider optical flats of 100 mm diameter are used to form the silvered liquid wedge. The optical system provides a wider cross section of the collimated monochromatic beam illuminating the interferometer. The wedge angle ε is kept in the range $5 \times 10^{-3} - 10^{-4}$ radians as it also determines the fringe spacing in the liquid region.

The multiple-beam fringes in the liquid region are straight lines parallel to the edge of the wedge with spacing $\Delta Z = \lambda/2n_L \tan \varepsilon$. As they cross the liquid/clad interface of a fibre having circular cross section, they follow elliptic shapes as previously derived mathematically. Figures 4.16, 4.17 and 4.18 show examples of microinterferograms of different fibres.

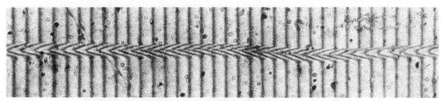

Figure 4.16 Multiple-beam Fizeau fringes of a natural fibre (camel hair) at reflection, $t_f = 40$ μm. (From Barakat *et al* 1975.)

Figure 4.17 Multiple-beam Fizeau fringes of a monomode step-index optical fibre at reflection, $t_c = 8$ μm.

Figure 4.18 Multiple-beam Fizeau fringes of a multimode graded-index optical fibre in transmission, $t_f = 125$ μm and $t_c = 50 \pm 1$ μm.

4.5.2 Opto-thermal properties of fibres

The thermal coefficient of refractive index dn/dT for a fibre is determined by forming Fizeau fringes for the same cross section of the fibre when immersed in a silvered liquid wedge at temperatures T_1, T_2, ... Then $(n_a)_{T_1}$ and $(n_a)_{T_2}$ are calculated from the fringe shift $\delta Z/\Delta Z$ for the two vibration directions, with the knowledge of n_L at the two temperatures. The mean value of dn/dT for acrylic fibres as reported by Barakat and El-Hennawi (1971) equals 3.3×10^{-4} $°C^{-1}$. The thermal coefficient dn/dT and bire-fringence of γ-irradiated Dralon fibres under vacuum were reported by Hamza and Mabrouk (1988), $dn_a^{\parallel}/dT$ and $dn_a^{\perp}/dT$ were found to be -9×10^{-4} $°C^{-1}$ and 7.5×10^{-4} $°C^{-1}$, the dose being 22.6 Mrad over a temperature range 26–31.5 $°C$.

4.5.3 Interferometric investigations of opto-mechanical properties of fibres

Synthetic fibres in the drawn or extended state show considerable optical and mechanical anisotropy. The degree of anisotropy in the drawn state is related to the amount of extension imposed. Optical anisotropy is con-venient for determining orientation in polymer films. Kuhn and Grün (1942) developed a theory yielding a relation between the molecular structure of a uniaxially oriented polymer and its optical anisotropy.

de Vries (1959) gave an analysis of the relationship between the birefringence and the draw ratio of synthetic fibres. Pinnock and Ward (1964) studied a series of polyethylene terephthalate fibres of different draw ratios. They measured the mechanical and optical properties of these fibres and attempted to develop both mechanical and optical anisotropy on a theoretical basis in terms of the orientation of the polymer molecules. Barakat and Hindeleh (1964b) reported the effect of stretch on refractive indices and birefringence of normal viscose rayon fibres. Hamza and Kabeel (1987) studied interferometrically the optical anisotropy in polypropylene fibres as a function of the draw ratio. Figures 4.19 and 4.20 show microinterferograms of multiple-beam Fizeau fringes in transmission for polypropylene fibres with draw ratios 3.0 and 4.0, respectively, for light vibrating parallel and perpendicular to the fibre axis. They reported the mean refractive indices $n_a^\parallel$, $n_a^\perp$, the core refractive indices $n_c^\parallel$, $n_c^\perp$ and the variation of the mean birefringence with the draw ratio. Figure 4.21 shows a schematic microstrain device for opto-mechanical investigations of fibres (see Hamza *et al* 1987).

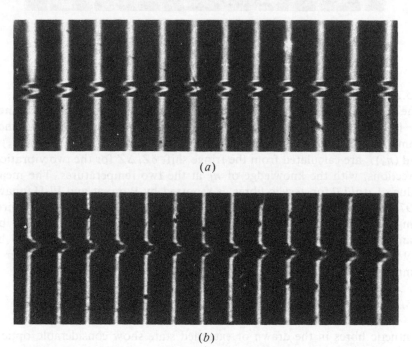

(a)

(b)

Figure 4.19 Microinterferograms of multiple-beam Fizeau fringes in transmission for a polypropylene fibre with draw ratio 3 for light vibrating parallel (*a*) and perpendicular (*b*) to the fibre axis. (From Hamza and Kabeel 1987.)

4.5.4   Dispersion properties of fibres

The variation of the mean refractive index of a fibre with wavelength, $dn_a/d\lambda$, is measured interferometrically either by applying multiple-beam Fizeau fringes or fringes of equal chromatic order. In both cases this leads to calculating the constants A and B of Cauchy's dispersion formula, namely $(n_a)_\lambda = A + B/\lambda^2$, when the dispersion of the fibre is normal over the specified range of wavelengths. When applying Fizeau fringes, monochromatic light at various wavelengths illuminates the silvered liquid interferometer with the fibre immersed, thus forming an interferogram for

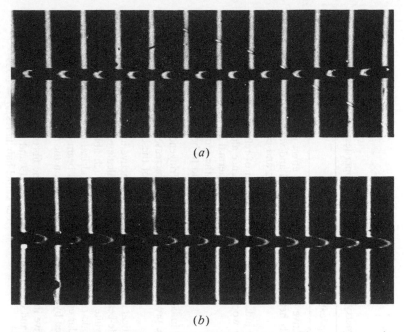

(*a*)

(*b*)

Figure 4.20 Microinterferograms of multiple-beam Fizeau fringes in transmission for a polypropylene fibre with draw ratio 4 for light vibrating parallel (*a*) and perpendicular (*b*) to the fibre axis. (From Hamza and Kabeel 1987.)

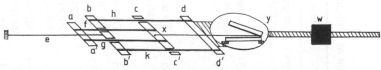

Figure 4.21 Schematic microstrain device. (From Hamza *et al* 1987.)
aá, movable bar; cć, sliding bar; f and g, two sliding rods; bb́dd, fixed frame; e, threaded rod; x, clamp for tightening one end of the tested fibre; w, movable weight to make the system stable; y, wedge interferometer.

each wavelength used. A mercury lamp, or a mercury–cadmium lamp, emits a suitable number of spectral lines, with the appropriate filters providing suitable sources of illumination. From the fringe shift $(\delta Z/\Delta Z)_\lambda$ for the two directions of light vibrations, $(n_a^{\|})_\lambda$ and $(n_a^{\perp})_\lambda$ and their variation with $1/\lambda^2$ are investigated. The constants A and B of Cauchy's formula are calculated for normal dispersion, i.e. when n_a holds a linear relation with λ^{-2}. Hamza and Mabrouk (1988) applied the Fizeau method to Dralon fibres, γ-irradiated with a dose of 22.559 Mrad, using Hg spectral lines in

Table 4.1 Application of multiple-beam interferometry to the study of fibre properties.

Author	Method	Object of study and application	Results
Faust (1952, 1954)	Multiple-beam inter-ferometry	Determination of refractive index variation within optically heterogeneous specimens	The skin effect in rayon fibres is discussed and values of $n^{\parallel}$ and $n^{\perp}$ for both skin and core are given
Barakat and Hindeleh (1964a)	Multiple-beam inter-ferometry	To determine refractive indices and birefringence of mohair wool fibres	Variation of refractive indices and birefringence along the fibre axis is given. Thermal coefficient of refractive index of the mohair fibre is determined and found to be $7.5 \times 10^{-5}\,°C^{-1}$
Barakat and Hindeleh (1964b)	Multiple-beam inter-ferometry	To determine refractive indices, birefringence and tensile properties of viscose rayon fibres	The birefringence of viscose rayon fibres is increased by increasing the tenacity of these fibres
Barakat (1971)	Multiple-beam inter-ferometry	Derivation of mathematical expression for the shape of multiple-beam Fizeau fringes and associated white light fringes of equal chromatic order crossing a fibre of circular cross section having a core surrounded by a skin	The refractive indices and birefringence can be calculated for both skin and core of such fibres. The optical power of a cylindrical fibre was calculated for a parallel beam of monochromatic light incident on the fibre
Barakat and El-Hennawi (1971), Barakat et al (1975)	Multiple-beam Fizeau fringes and the white fringes of equal chromatic order	Measurement of refractive indices and birefringence of acrylic and camel-hair fibres	For acrylic fibres, $n^{\parallel} = 1.518$, $n^{\perp} = 1.519$ and $\Delta n = -0.001$ at 35 °C. For camel-hair fibres, $n^{\parallel} = 1.559$, $n^{\perp} = 1.546$ and $\Delta n = 0.013$ at 21.5 °C

Reference	Method	Subject of study	Results
Hindeleh (1978), Hamza and Sokkar (1981)	Multiple-beam Fizeau fringes	Study of the optical anisotropy in cotton fibres	The values of the mean refractive indices $n_a^{\parallel}$ and $n_a^{\perp}$ and birefringence of cotton fibres differ for different varieties
Krishna Iyer et al (1969)	White light fringes of equal chromatic order		The refractive indices of each layer of the fibre and their variations with wavelength of light were determined
Hamza et al (1980a,b)	Immersion and multiple-beam Fizeau methods	Investigation of the difference in the mean orientation of skin and core, for polyethylene and polypropylene fibres.	The optical properties of multiple-skin fibres of elliptical, rectangular, kidney and dog-bone cross sections are given
El-Niklawy and Fouda (1980a,b), Fouda and El-Niklawy (1981), Fouda et al (1981), Hamza et al (1982)	Fizeau method	Derivation of mathematical expressions for the shape of multiple-beam Fizeau fringes and their application to determine refractive indices of multiple-skin fibres	The results are in good agreement with those obtained from the double-beam microinterferometric method
Barakat and El-Hennawi (1971), Hamza and Abd El-Kader (1983)	Fizeau method	Description of a method suitable for evaluating small birefringence in fibres and its application to acrylic and cuprammonium fibres	
Sokkar and Shahin (1985), Hamza et al (1984,1985a,b,c)	Double-beam and multiple-beam microinterferometry	Determination of the optical anisotropy of fibres with irregular transverse sections.	Accurate results are obtained when considering the area under the interference fringe shift represented by the path difference integrated across the fibre. Values of refractive indices and birefringence for the skin and core of a fibre having irregular transverse sections are given

the visible, vibrating parallel to the fibre axis, and reported $A = 1.5149$ and $B = 15.53 \times 10^2$ (nm)^2.

4.5.5 Fringes of equal chromatic order

The multiple-beam monochromatic system in transmission produced by the method described in the preceeding section is projected on the slit of a prism or grating spectrograph and the monochromatic source is replaced by a (Pointolite) white light source. By adjusting the edge of the wedge parallel to the slit of the spectrograph, white light fringes of equal chromatic order are formed and appear as vertical straight lines in the liquid part of the interferometer (Tolansky 1960). As these fringes cross the fibre, their shift varies with the wavelength of light and with the direction of light vibration.

Figure 4.3 shows the set-up for producing fringes of equal chromatic order (FECO). The fringe suffering no shift indicates that n_a is equal to n_L for the wavelength λ_1. When the temperature of the experiment is slightly varied, equality occurs at another close wavelength λ_2. A family of dispersion curves for the liquid used over the specified range of wavelengths and at temperatures between T_1 and T_2 are constructed using a thermostatically controlled refractometer and illuminated by monochromatic light. From these curves n_a of the fibre at the required wavelength can be determined. The constants of Cauchy's dispersion formula can also be calculated when the fibre dispersion is normal. Barakat and Hindeleh (1964b) applied the above method and reported the values of the constants of Cauchy's formula for normal viscose rayon fibres to be $A = 1.5391$ and $B = 266.666$ nm^2. FECO fringes were used to determine the birefringence of acrylic fibres by Barakat and El-Hennawi (1971).

A comprehensive survey of applications of multiple-beam Fizeau and FECO fringes was published by Hamza (1986). It is quoted here in table 4.1, and recommended for consultation for more detailed investigation in the literature.

References

Barakat N 1957 *Proc. Phys. Soc* B **IXX** 220
—— 1971 *Textile Res. J.* **41** 167
Barakat N and El-Hennawi H A 1971 *Textile Res. J.* **41** 391
Barakat N, Hamza A A and Fouda I 1975 *Egypt. J. Phys.* **6** 91
Barakat N, Hamza A A and Goneid A S 1985 *Appl. Opt.* **24** 4383
Barakat N and Hindeleh A M 1964a *Textile Res. J.* **34** 357
—— 1964b *Textile Res. J.* **34** 581
Barakat N and Mokhtar S 1963 *J. Opt. Soc. Am.* **53** 159
Brossel J 1947 *Proc. Phys. Soc.* **59** 224
El-Hennawi H A 1988a *Egypt. J. Phys.* in press
—— 1988b *Egypt. J. Phys.* in press

El-Hennawi H A 1988c *Egypt. J. Phys.* in press
El-Nicklawy M M and Fouda I M 1980a *J. Textile Inst.* **71** 252
—— 1980b *J. Textile Inst.* **71** 257
Faust R C 1952 *Proc. Phys. Soc.* B **65** 48
—— 1954 *Proc. Phys. Soc.* B **67** 138
Feussner W 1927 *Gehrckés Handbook der Physik Optik* vol. 1
Fouda I M and El-Nicklawy M M 1981 *Acta Phys. Polon.* A **59** 95
Fouda I M, Hamza A A, El-Nicklawy M M and El-Farahaty K A 1981 *Textile. Res. J.* **51** 355
Gloge D and Marcatili E A J 1973 *Bell Syst. Tech. J.* **52** 1563
Hamza A A 1980 *Textile Res. J.* **50** 731
—— 1986 *J. Microsc.* **142** 35
Hamza A A and Abd El-Kader H I 1983 *Textile Res. J.* **53** 205
Hamza A A, Fouda I M and El-Farahaty K A 1982 *Acta Phys. Polon.* A **61** 129
Hamza A A, Fouda I M, El-Farahaty K A and Badawy Y K M 1980a *Textile Res. J.* **50** 592
—— 1980b *Acta Phys. Polon.* A **58** 651
Hamza A A, Fouda I M, El-Farahaty K A and Helaly S A 1987 *Polym. Test.* **7** 329
Hamza A A, Fouda I M, Hashish A H and El-Farahaty K A 1984 *Textile Res. J.* **54** 802
Hamza A A and Kabeel M A 1986 *J. Phys. D: Appl. Phys.* **19** 1175
—— 1987 *J. Phys. D: Appl. Phys.* **20** 963
Hamza A A and Mabrouk M A 1988 *Radiat. Phys. Chem.* **32** 645
Hamza A A and Sokkar T Z N 1981 *Textile Res. J.* **51** 485
Hamza A A, Sokkar T Z N and Kabeel M A 1985a *J. Phys. D: Appl. Phys.* **18** 1773
—— 1985b *J. Phys. D: Appl. Phys.* **18** 2321
—— 1986 *J. Phys. D: Appl. Phys.* **19** L19
Hamza A A, Sokkar T Z N and Shahin M M 1985c *J. Microsc.* **137** 85
Hindeleh A M 1978 *J. Phys. D: Appl. Phys.* **11** 2335
Krishna Iyer K R, Neelakantan P and Radhakrishnan T 1969 *J. Appl. Polym. Sci.* **7** 983
Kuhn W and Grün F 1942 *Kolloid Z.* **101** 248
Marhic M E, Ho P S and Epstein M 1975 *Appl. Phys. Lett.* **26** 574
Mokhtar S 1964 *PhD Thesis* Ain Shams University, Egypt
Saunders M J and Gardner W B 1977 *Appl. Opt.* **16** 2368
Simmens S C 1958 *Nature* **181** 1260
Sokkar T Z N and Shahin M M 1985 *Textile Res. J.* **55** 139
Pinnock P R and Ward I M 1964 *Br. J. Appl. Phys.* **15** 1559
Tolansky S 1948 *Muliple-Beam Interferometry* (Oxford: Clarendon)
—— 1960 *Surface Microtopography* (London: Longmans, Green)
de Vries H 1959 *J. Polym. Sci.* **34** 761

5 Interferometric Determination of Fibre Surface Topography

Interferometric techniques have been used for the surface examinations of materials to detect and evaluate fine detail. Both multiple-beam interferometric techniques and two-beam interference microscopes have been used. Tolansky (1948, 1952, 1960) and his co-workers made an extensive study of the topography of crystal and metal surfaces. They applied multiple-beam localised interference systems in transmission and at reflection.

5.1 Multiple-beam localised interference systems in transmission applied to surface topography

The optical system of multiple-beam interferometry developed by Tolansky to investigate the surface topography of transparent objects was described in Chapter 4. It is the same optical arrangement used to form multiple-beam Fizeau fringes in transmission to determine the refractive indices of fibres, except in the mode of construction of the interferometer. To investigate the optical properties of a fibrous material the fibre is immersed in a silver liquid wedge, while to examine the surface topography of transparent objects, including fibres, multiple-beam Fizeau fringes are formed resulting from rays reflected at the surface under examination and a reference optical flat. Such a system of interference is localised at a definite position in space very close to the interferometer. Using thermal evaporation in a pressure of $< 10^{-5}$ Torr the two surfaces are coated with highly reflecting, partially transparent silver layers. They are then held in a 'Jig', forming a silvered air interferometer enclosing an air film; the silvered cylindrical surface and the silvered optical flat (figure 5.1) being the two components of the inter-

98

ferometer. Multiple-beam Fizeau fringes in transmission are formed. They appear as sharp bright straight-line fringes parallel to the fibre axis. The fringe spacing decreases as one moves away from the centre. Fine details of the surface topography of the fibre appear as deviations of the fringes from the regular pattern. They are due to elevations or depressions on the fibre surface. Also, the fringe spacings of successive fringes allow the determination of the fibre radius for different cross sections. This is shown in figure 5.2.

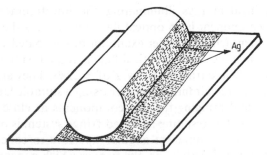

Figure 5.1   An interferometer for determining surface topography of a fibre forming fringes in transmission.

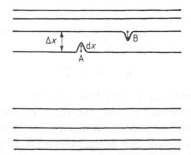

Figure 5.2   The shape of localised multiple-beam fringes, straight sharp bright fringes in transmission and sharp dark fringes at reflection, parallel to the fibre axis. The fringe spacing decreases as one moves away from the centre, governed by the curvature of the fibre surface. The fringe shift A represents a hill or elevation on the fibre surface as it is towards longer interferometric gap thickness t_g, while the fringe shift B results from a valley or a depression.

5.2   Multiple-beam localised interference systems at reflection applied to surface topography

Multiple-beam Fizeau fringes at reflection have been applied by Tolansky and his co-workers to examine the surface topography of a variety of crystals, including mica, selenite and quartz for the identification of surface characteristic features and step heights and silicon carbide for the examination of growth features and indentation for hardness evaluation of metals. The optical arrangement is similar to that presented in Chapter 4 except in the mode of construction of the interferometer.

We will discuss the case where the interferometer is composed of the surface under examination fixed to a substrate, such that, by the aid of a microscope, the area selected on the surface is rendered horizontal with respect to the incident rays illuminating the interferometer. The other interferometer component is an optical flat, which faces the incident light. The first component (surface under examination) is coated with an opaque layer of silver, while the second (optical flat) is coated with a silver layer of reflectivity $R \approx 70\%$ and transmissivity T of $\approx 22\%$. They are then brought close, to form a silvered air interferometer using a suitable 'Jig', enclosing a thin layer of air and bringing the two components as close as possible as shown in figure 5.3. The resulting localised fringe system is observed with a microscope objective of sufficient angular aperture to collect all effective beams. If λ is the wavelength in air, the fringes which appear as sharp dark lines on a bright background are contours of the unknown surface defined by planes parallel to the optical flat at intervals of $\lambda/2$. The fringe spacing is governed by the mutual inclination of the two components of the interferometer. The topography of the unknown surface is evaluated quantitatively by measuring the fringe lateral shift dx corresponding to an elevation h or depression on the fibre surface

$$h = \Delta m(\lambda/2) \qquad (5.1)$$

where $\Delta m = dx/\Delta x$ and Δx is the fringe spacing as shown in figure 5.2.

Thus h can be determined if Δm, the ratio of the fringe shift and the adjacent fringe spacing, is known. When the reflectivity is high, surface features of very small height can be measured. A fringe shift equal to the fringe half width, which is easily measurable, corresponds to $h = \lambda/2F$ with $\lambda = 5500$ Å and finesse $F = 40$; h is only about 70 Å (see Born and Wolf 1980). Whether a surface feature is an elevation or a depression can be decided by observing the direction of motion of the fringes when the separation of the two components of the interferometer is altered. This is applicable to interference systems both in transmission and at reflection. When applying interference systems in transmission, it is also possible to distinguish between a hill and a valley on the surface under examination by using more than one wavelength. The accuracy in measurement depends on the quality of the fringes. Holden (1949) dealt with some of the parameters governing multiple-beam Fizeau fringes at reflection, thus rendering the system more useful for applications. He has shown that with silver coatings of high reflectivity and low absorption, the fringes at reflection are narrower than the corresponding bright fringes in transmitted light, but the intensity at the minima of the reflected pattern is critically dependent on the absorption of the layer on the entrance surface. In fact the optical phase properties of this layer, namely its F value where $F = 2\gamma - (\beta_1 + \beta_2)$, are what governs the intensity distribution in multiple-beam Fizeau fringes at reflection (γ is the change of phase of light in transmission through the layer, β_1 and β_2 are the change of phase of reflected light at the air layer and substrate layer, as explained in Chapter 2).

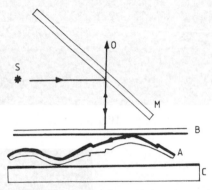

Figure 5.3   An interferometer for determining the surface topography of mica. A is a freshly cleaved mica film silvered with an opaque layer from one side; B, a reference silvered optical flat. They are brought as close as possible, enclosing a thin layer of air. Multiple-beam fringes are formed at reflection. S is a parallel beam of monochromatic light, M a beam splitter, C a substrate and O leads to the microscope objective and camera.

Multiple-beam fringes of equal chromatic order have been used to examine the surface topography of crystals (see Tolansky 1960). The optical arrangement used was described in Chapter 4.

As illustrations of the comparison between the quantitative estimation of fine details on the surface of objects when applying multiple-beam and two-beam localised interference systems, figures 5.4(a, b) show interferograms of a spherical surface of a transparent object, using the multiple-beam technique in transmission and at reflection respectively, while figure 5.4(c) shows an interferogram for two-beam fringes. The multiple-beam fringes are extremely sharp and fine and reveal secondary microstructure on the surface of the object, which is almost completely lost in the broad two-beam fringes.

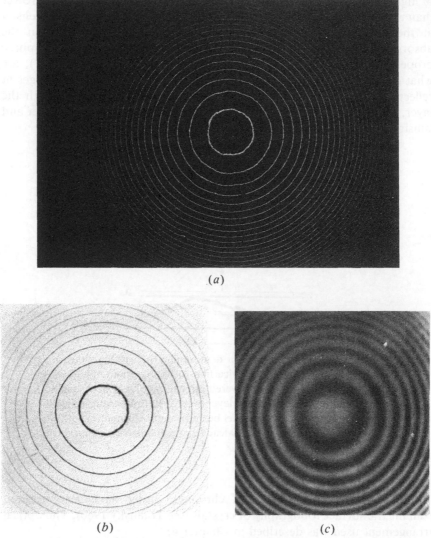

(a)

(b)   (c)

Figure 5.4 (*a*) Multiple-beam interferogram in transmission. (*b*) Multiple-beam interferogram at reflection. (*c*) Two-beam interferogram for the same spherical surface.

5.3   The application of interference microscopy to the study of fibre surface topography

Interference microscopy has been applied to the study of surface topography of textile fibres (Skretchly 1954, Howell and Mazur 1953, Simmens

1954) and optical fibres (Barakat *et al* 1986). In one of these studies, the techniques developed by Tolansky (1952) for the observation of crystal growth features were applied by Skretchly (1954) to the examination of the surface topography of some keratin fibres. The tested fibre is pretreated in a solution of Canada balsam dissolved in benzene. After drying, this solution leaves a thin film of about two micrometres thick, with a base that follows the film contour; the surface of the film remains smooth and curved. The fibre is then mounted on a microscope slide and fixed in its position by an adhesive. A microscope illuminated by reflected monochromatic light is used to observe the fringes resulting from the interference of light reflected at the two Canada balsam surfaces, the outer surface acting as the reference surface. As reflected light is used in this technique, the lens action of the fibre has no influence on the interference effect.

A technique used by Howell and Mazur (1953) involved the production of Newton's rings for examining contact areas. Their work aimed at the study of fibre friction.

Simmens (1954) described a simple interferometric method for the examination of the surface topography of fibres and filaments. Figure 5.5 shows the optical set-up used, which is similar to the optical arrangement used to observe Newton's rings. With this technique, interference occurs between light reflected at the filament and a reference plate, the filament being observed to be contoured by a system of interference fringes. The difference in surface heights which are observed corresponds to lines of constant heights.

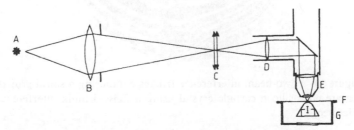

Figure 5.5 The optical set-up for fibre surface investigations. (From Simmens 1954.)
A, monochromatic light source; B, diaphragm and collecting lens focused to project an image of A into C; C, two-way adjustable slit; D, vertical illuminator unit and condenser lens; E, microscope objective lens; F, support mounted on substage condenser unit; I, filament.

It is to be noted that all optical arrangements described so far form localised interference fringes (whether they are in transmission or at reflection, or a multiple-beam or two-beam intensity distribution) close to the interferometer. The fringe system is then magnified and recorded. A low-power objective is needed, with a depth of focus long enough to bring the fringes in focus over an appreciable area.

The Zeiss–Linnik interference microscope (ZLIM) has been used to determine the surface topography of objects; a detailed account of the optical components, the passage of light and the experimental procedure used to form two-beam interference fringes with high magnification is given in Chapter 7. Barakat (1961) used the ZLIM to investigate growth features on the surface of a silicon carbide crystal. Figure 5.6 shows a microinterferogram of a spiral growth feature on one surface of a silicon carbide crystal containing a polytype 27 H (see Mitchell *et al* 1958). The optical arrangement is so adjusted to allow the two-beam intensity distribution to contour the spiral, thus giving its height as intensity difference as well as visualising it.

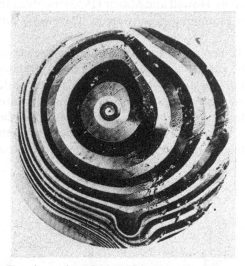

Figure 5.6   Two-beam interference fringes contouring a spiral growth feature on a silicon carbide crystal using a Zeiss–Linnik interference microscope.

Barakat *et al* (1986) used a Zeiss–Linnik interference microscope to determine the surface topography of fusion-spliced optical fibres. The method aimed to control and examine the quality of the splicing process, since the practical implementation of optical fibre communication systems requires the use of interconnection devices such as splices or connectors. The loss introduced by splices and connectors is an important factor in the design of a fibre-optic system since it can contribute significantly to the losses of a multikilometre communication link. A graded-index fibre of 50 μm core and 125 μm clad diameter was chosen and a Siecor M67 fusion splicer was used. The ends of the fibres to be spliced were stripped of their coating, cleaned and a fibre cutter was used for a perfect cut of the fibre end. Prefusion was used to remove impurities at the ends of fibres, then they were brought together and spliced. The Zeiss–Linnik interference microscope was used to determine the surface topography of the fusion-spliced graded-index optical fibres. Interference takes place between two reflected beams, one from a reference plane mirror and the other reflected from the surface of the fibre under test. The reflectivity of the reference mirror was chosen to be closest to that of the fibre. The monochromatic source used was a thallium lamp emitting at $\lambda = 535$ nm. A white light source replaced the monochromatic source to form low-order white light fringes. The spliced optical fibre was placed in a 'Jig' to render that part of its surface selected by the objective perpendicular to the incident light. Figure 5.7 shows an interferogram of the spliced part of a graded-index optical fibre using the Zeiss–Linnik interference microscope at $\lambda = 535$ nm. Two closed-fringe systems appear at B and C on both sides of splicing point A, due to buckling of the fibre material resulting from the splicing process. The height of the buckling h was deduced from equation (5.1) in terms of Δm the number of interference fringes enclosed between A and B or A and C, and an additional fraction at the centre of the fringe system. The buckling heights at B and C are 6.7 and 2.9 μm, respectively, where B and C are 300 μm apart. Figure 5.8 shows an interferogram of part of the spliced fibre starting from the splicing point at A, covering the buckling at B and extending to the unperturbed part of the optical fibre, where it shows straight fringes parallel to the fibre axis.

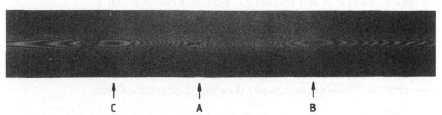

Figure 5.7 An interferogram of the spliced part of a graded-index optical fibre. (From Barakat *et al* 1986.)

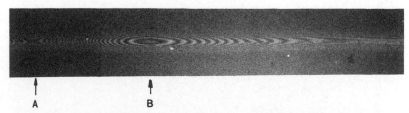

Figure 5.8 An interferogram of a spliced optical fibre, the splicing point is at A. A closed fringe system appears at B indicating the presence of buckling. (From Barakat *et al* 1986.)

A comparison between the average height of buckling formed due to splicing and the corresponding loss in power transmitted has been experimentally achieved. It indicates that the higher the value of buckling height, the greater the power loss, with the criterion that the size of buckling must be within 1 μm. The straight fringe system parallel to the fibre axis, formed by the unperturbed cylindrical part of the fibre, allows determination of the fibre diameter d along the fibre using the formula

$$d = (D_m^2/m)(1/2\lambda) \tag{5.2}$$

where D_m denotes the distance between any two straight fringes of the same order of interference m divided by the magnification on the photographic plate. The fibre diameter has been evaluated from the slope of the straight line between D_m^2 and m. The magnitude of the fibre diameter evaluated at a point is accurate to ± 1 μm with an estimated average of 126 μm.

References

Barakat N 1961 *Zeiss-Mitteilungen (Heft. FRG.)* 6 325
Barakat N, El-Hennawi H A, Medhat M, Sobie M A and El-Diasti F 1986 *Appl. Opt.* 25 3466
Born M and Wolf E 1980 *Principles of Optics* (Oxford: Pergamon) p357
Holden J 1949 *Proc. Phys. Soc.* B 62 405
Howell H G and Mazur J 1953 *J. Textile Inst.* 44 T59
Mitchell P, Barakat N and El-Shazly E 1958 *Z. Kristallogr.* III 1
Simmens S C 1954 *J. Textile Inst.* 45 T569
Skretchly A 1954 *J. Textile Inst.* 45 T78
Tolansky S 1948 *Multiple-Beam Interferometry of Surfaces and Films* (Oxford: Clarendon)
—— 1952 *Nature* 170 4315
—— 1960 *Surface Microtopography* (London: Longmans, Green)

6 The Effect of Irradiation on the Optical Properties of Fibres

The present chapter deals with the effect of irradiation on the optical properties of fibres. It constitutes two sections; §6.1 is concerned with the effect of γ and neutron irradiation on the optical properties of optical fibres, namely optical absorption, while §6.2 presents the effect of γ irradiation on the refractive indices and birefringence of textile synthetic fibres and optical fibres.

6.1 The effect of γ and neutron irradiation on the optical absorption of optical fibre waveguides

When optical fibre waveguides are exposed to nuclear radiation, the optical absorption of the fibre increases. Such induced absorption causes induced loss and an increase in attenuation. This radiation-induced absorption is of concern when optical communication systems are employed in radiation environments. Numerous studies have been conducted to investigate the behaviour of fibres during and after irradiation and to explain the mechanisms responsible for the radiation-induced loss (Maurer *et al* 1973). With the production of low-loss waveguides and newer core and cladding compositions, studies of radiation damage in both polymer-clad silica (PCS) by Friebele *et al* (1978b) and doped silica and plastic fibres (Friebele *et al* 1978a) have been undertaken. The availability of long lengths of low- and moderate-loss fibres made it possible to study the growth and decay of the radiation-induced attenuation from doses as low as 1 rad to as high as 10^6 rad.

Optical communication systems, particularly over distances less than

2 km, operate in the near-infrared at about 850 nm, where light emitting diodes (LED) and injection lasers are available; most characterisations of low-loss fibres have been carried out at about 0.82 μm. Friebele *et al* (1979) extended their investigations to the behaviour of optical fibres near 1.3 μm. Their results at 0.82 μm have revealed that the damage induced in PCS fibres saturates with increasing dose so that the induced loss at low dose is more than two orders of magnitude greater than that anticipated on the basis of high dose measurements of short lengths of fibres or bulks. It was also observed that (i) PCS fibres with low OH content were more susceptible to radiation damage at 0.82 μm than those with high OH content, (ii) a very large transient absorption was observed in germanium-doped silica core fibres and (iii) spectral measurements of absorption spectra of fibres between 0.4 and 1.0 μm have indicated a general decrease in the radiation-induced absorption upon going to longer wavelengths. Characterisation of the radiation damage in low-loss fibres near the wavelength $\lambda = 1.3$ μm is becoming essential for optical communication systems using lasers emitting at this wavelength because of the remarkably low material dispersion of optical fibres. Concerning the physics of the damage mechanisms responsible for the observed radiation-induced absorption, two types of damage are considered.

(*a*) Damage in fibres by γ rays. The γ rays interact with glasses principally by forcing the electrons to leave their normal positions and move through the glass network. The primary consequence is an increase in the absorption coefficient in the ultraviolet and visible-near-infrared range. Sigel and Evans (1974) studied the γ-induced damage in fibres and found that γ-induced losses depend strongly on the fibre composition and vary from 10^{-4} dB km^{-1} rad^{-1} for bulk Suprasil SiO$_2$ to 5 dB km^{-1} rad^{-1} for Corning fibre at 8000 Å. Thus pure fused silica is extremely resistant to radiation while Corning 5010 is quite susceptible to it.

(*b*) Damage in fibres by neutrons. In solids neutrons interact principally with the nuclei (see Shah 1975). Neutron radiation, therefore, results not only in increased absorption losses but also in structural changes that lead to changes in density, refractive index, rotary power, birefringence and thermal conductivity. Maurer *et al* (1973) irradiated high-silica-glass multi-mode fibre waveguides with 14 MeV neutrons using doses as high as 1.4×10^{12} neutrons/cm^2. They concluded that the neutron-induced loss varies roughly linearly with the total dose and is less than 1.5×10^{-11} (dB km^{-1})/(neutrons/cm^2) in the 8000 to 12 000 Å region.

For permanent spectral measurements Friebele *et al* (1979) used 10–30 m of the optical fibre irradiated with a Co^{60} source of 10^5 rad (Si) and the optical absorption from 0.4 to 1.7 μm was measured prior to and one hour after the irradiation. One metre of the fibre was exposed to a 3 ns, 3700 rad, 0.5 MeV pulsed electron irradiation. The fibres used by Friebele *et al* were graded-index, and similar results have been obtained for step-index fibres. Their results showed that in addition to broad radiation-induced absorption bands in the ultraviolet and infrared, there is an increase in the intensity of the OH overtone and combination bands at 0.95, 1.23 and 1.3 μm.

6.2 The effect of γ irradiation on the refractive indices and birefringence of optical and synthetic textile fibres

6.2.1 The effect of γ irradiation on the refractive indices of optical fibres

Bertolotti *et al* (1979, 1980a,b) applied a two-beam interference technique to study small changes in refractive index of γ-irradiated optical fibres. They reported that sensible changes in both refractive indices and dimensions of optical fibres occur even in the case of relatively low radiation doses of γ rays, e.g. 1 krad. Moreover, the effects anneal out at room temperature in a few days. In the following, a description is given of the techniques used by Bertolotti and his co-workers. Figure 6.1 shows a two-beam interferometric system that allows formation of interference fringes, resulting from a beam crossing the fibre placed in arm A and another reference beam propagating along arm B, containing the reference fibre specimen, of a Mach–Zehnder interferometer. It is a single-pass interferometer. The fringes are focused on the photographic plate P through the optics O. This method was used to determine the changes in the refractive index and thickness of step-index optical fibres, irradiated with 1 krad of γ rays emitted from a Co[60] source. Bertolotti *et al* reported relative changes in the core and cladding radii of 3.9 and 1.8%, respectively, and in the core and cladding refractive indices relative changes of 2.8 and 2.26%, respectively.

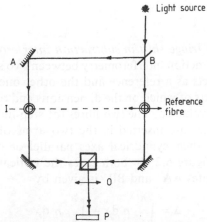

Figure 6.1 A two-beam interferometric system applied to study small changes in refractive indices of γ-irradiated optical fibres. (From Bertolotti *et al* 1980b.)

A subtraction interferometric method which proved to be useful for the study of small changes in refractive indices and dimensions of optical fibres was described by Bertolotti *et al* (1980a). This method is shown schematically in figure 6.2. Two fibres are inserted in the two arms of a Mach–Zehnder interferometer with their cylindrical axis parallel (position *a*) or orthogonal (position *b*).

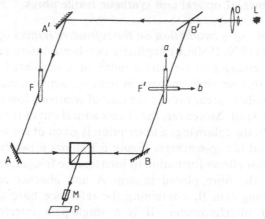

Figure 6.2 A schematic representation of the subtraction interferometric method applied to optical fibres. L is a light source, AA′ and BB′ are the components of a Mach–Zehnder interferometer, M is the microscope optics and P is a photographic plate. The two fibres are mounted with the fibre F′ in position *a* so as to have both axes of the two specimens parallel to each other or in position *b* so that the axes are at right angles with respect to each other. (From Bertolotti *et al* 1980a.)

Expressions for the fringe shift in subtraction interferometry of fibres
Let us consider subtraction interferometry between two samples of the same fibre; one sample acts as a reference and the other one has suffered some irradiation, the effects of which on the dimensions and refractive indices are to be detected and evaluated. The two fibres (of step-index type, composed of core and cladding) are inserted in the two arms of a Mach–Zehnder interferometer with their cylindrical axes parallel or orthogonal to each other as shown in figure 6.2. In both cases the optical path length (OPL) between the two routes AA′ and BB′ is given by

$$\Delta = \int_A^{A'} n \, dx - \int_B^{B'} n \, dx.$$

By considering small index variations and treating the fibre as a purely phase object, Δ is seen as a fringe shift.

6.2.2 Effect of γ irradiation on the refractive indices and birefringence of textile synthetic fibres

Hamza *et al* (1986) applied multiple-beam Fizeau fringes to study the effect of γ irradiation on some optical properties of polymeric fibres. The irradiation process was performed in air. Hamza and Mabrouk (1988) applied multiple-beam interference systems in transmission and at reflection to determine the mean refractive indices and mean birefringence of γ-irradiated Dralon fibres; the irradiation of the polymeric material took place under reduced pressure of 1.5×10^{-4} Torr. The samples were packed in glass test-tubes closed under vacuum, then exposed to the γ-irradiation source, a Co^{60} Gamma 3500 irradiation unit. The central dose rate was up to 2×10^{6} rad h^{-1}. Irradiation was carried out over a period of 96–573 h and the dose rate was maintained at 23.71 ± 0.27 rad s^{-1}. Figure 6.3 shows microinterferograms of multiple-beam Fizeau fringes at reflection crossing a Dralon fibre irradiated with a dose $r = 22.559$ Mrad for light vibrating parallel and perpendicular to the fibre axis, respectively, λ being 546.1 nm. One notices that the area F enclosed under the fringe shift, which enters in the calculation of the mean refractive index n_a according to the relation

$$n_a = n_L + \frac{F}{2A} \frac{\lambda}{h}$$

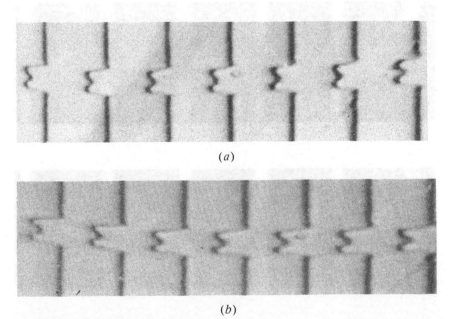

(*a*)

(*b*)

Figure 6.3 Microinterferograms of multiple-beam Fizeau fringes at reflection crossing a Dralon fibre γ-irradiated by 22.559 Mrad when monochromatic light vibrates parallel (*a*) and perpendicular (*b*) to the fibre axis. (From Hamza and Mabrouk 1988.)

as explained in Chapter 4 (p75), where A is the transverse sectional area of the fibre and h the interfringe spacing, is more for light vibrating perpendicular to the fibre axis. This indicates that $n_a^\perp$ is greater than $n_a^\parallel$ for the same dose. The values given by this irradiated fibre are $n_a^\parallel = 1.5178$, $n_a^\perp = 1.5202$ and birefringence $\Delta n_a = -2.4 \times 10^{-3}$. Figure 6.4 shows multiple-beam Fizeau fringes in transmission crossing the Dralon fibre irradiated with a 49.147 Mrad dose of γ radiation for light vibrating parallel and perpendicular to the fibre axis, respectively, λ being 546.1 nm. One notices from figures 6.3 and 6.4 that the area F enclosed under the fringe shifts, increases with increasing dose. Hamza and Mabrouk (1988) suggested empirical formulae to correlate the mean refractive index and birefringence with the dose r in rad, over the range of 0–49.147 Mrad, in the form

$$n_a = n_0 \exp(ar^{1/3})$$

where $n_0^\parallel = 1.5122$ and $a = 1.344 \times 10^{-5} \, \mathrm{rad}^{-1/3}$ for $n_a^\parallel$ and $n_0^\perp = 1.5164$ and $a = 9.08 \times 10^{-6} \, \mathrm{rad}^{-1/3}$ for $n_a^\perp$. When dealing with the mean birefringence Δn_a, they suggested the formulae $\Delta n_a = \Delta n_0 \exp(-ar^{1/2})$ with $\Delta n_0 = -4.2 \times 10^{-3}$ and $a = 1.285 \times 10^{-4} \, \mathrm{rad}^{-1/2}$.

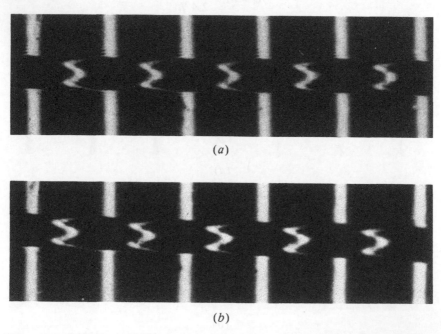

(a)

(b)

Figure 6.4 Microinterferogram of multiple-beam Fizeau fringes in transmission crossing a Dralon fibre γ-irradiated by 49.147 Mrad when monochromatic light vibrates parallel (a) and perpendicular (b) to the fibre axis. (From Hamza and Mabrouk 1988.)

For the case of Dralon fibres γ-irradiated in air, the same empirical formulae apply, but $n_0^\| = 1.5120$ and $a = 2.03 \times 10^{-5}$ rad$^{-1/3}$ for $n_a^\|$ and $n_0^\perp = 1.5162$ and $a = 1.62 \times 10^{-5}$ rad$^{-1/3}$ for $n_a^\perp$.

Barakat *et al* (1989) applied multiple-beam Fizeau fringes in transmission to determine the refractive indices $n_f^\|$, $n_f^\perp$ and the birefringence Δn for γ-irradiated blue Cashmeline fibres. For the wavelength $\lambda = 546.1$ nm, $n_f^\|$, $n_f^\perp$ and Δn were found to be 1.5118, 1.5145 and -0.0026, respectively, for these fibres before irradiation. They reported that the result of γ irradiation was a decrease in the birefringence as the dose increases. The two curves of $n_f^\|$ versus the dose r and $n_f^\perp$ versus r intersect at a dose of $r = 185$ Mrad. This indicates that the fibre behaviour is close to being isotropic. Further irradiation causes the two curves to diverge but with opposite sign for the birefringence, $n_f^\|$ is greater than $n_f^\perp$.

References

Barakat N, El-Hennawi H A, El-Okr M and Sharaf F 1989 *J. Phys. D: Appl. Phys.* **22** 786
Bertolotti M, Ferrari A and Scudieri F 1979 *Radiat. Eff. Lett.* **43** 177
—— 1980a *Opt. Acta* **27** 1143
Bertolotti M, Ferrari A, Scudieri F and Serra A 1980b *Appl. Opt.* **19** 1501
Friebele E J, Gingerich M E and Sigel Jr G H 1978a *Appl. Phys. Lett.* **32** 619
Friebele E J, Sigel Jr G H and Gingerich M E 1979 *Fibre Optics* ed. B Bendow and S Mitra (London: Plenum) p355
Friebele E J, Sigel Jr G H, Jaeger R E and Gingerich M E 1978b *Appl. Phys. Lett.* **32** 95
Hamza A A, Ghander A M, Oraby A H, Mabrouk M A and Guthrie J T 1986 *J. Phys. D: Appl. Phys.* **19** 2443
Hamza A A and Mabrouk M A 1988 *Radiat. Phys. Chem.* **32** 654
Maurer R D, Schiel E J, Kronenberg S and Lux R A 1973 *Appl. Opt.* **12** 2023
Shah J 1975 *Bell Syst. Tech. J.* **54** 1208
Sigel Jr G H and Evans B D 1974 *Appl. Phys. Lett.* **24** 410

7 Interference Microscopes

Interference microscopes are modified microscopes such that whilst an object (either opaque or transparent) is under view, the instrument simultaneously behaves as an interferometer (see Tolansky 1973). In other words an interference microscope combines the two functions of the interferometer and the microscope in a single instrument. This modification allows one to draw additional useful information to that obtainable using a conventional microscope. There are many types of optical interferometers which can be adapted to optical microscopes and many commercial models are available.

7.1 Fundamentals of interference microscopy

Objects in optical microscopy can be classified as amplitude or phase objects. Amplitude objects vary in their light absorption with respect to the medium that surrounds them, thus exhibiting a certain amount of natural contrast. The human eye and the photographic plate are sensitive to changes in intensity of light. In interference microscopy, phase objects are of special interest. These objects produce no variations in the light being absorbed and differ from the surrounding medium merely by their refractive indices or thicknesses (nt). The phase object is placed in an interferometer and alters the optical path of the light beams passing through it.

Figure 7.1 is a diagram showing the principle of the two-beam interference microscope (see Françon 1961). Light beam SM from the condenser is split into two beams at M by one of the interferometer's elements. The light beam MON passes through the phase object O. This object may be a fibre. The beam MBN by-passes this object. At N the two beams are recombined by the other interferometer element and pass into the microscope to produce an interference effect. The intensity of the beam NS' is determined by the interference of the two beams MON and MBN and depends upon the refractive index and the thickness of the object O. The phase difference δ between the two waves W_1 and W_2 is equal to $(2\pi/\lambda)\Delta$, Δ being their path difference. This is regulated by means of the interferometer.

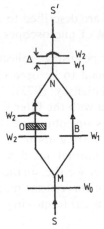

Figure 7.1 Principle of the two-beam interference microscope and the resulting wave surfaces.

The intensity distribution in the resulting fringe system

$$I = I_0 \cos^2 \delta/2$$

$$= I_0 \cos^2 \frac{\pi}{\lambda} \Delta$$

when $\Delta = 0$, $\delta = 0$ and $I = I_0$. The two waves are in phase in all regions other than that deformed by the phase object O. The distribution of intensity within the two-beam fringe pattern follows the $\cos^2$ law. The bright and dark fringe widths are equal, and represent half the order separation between consecutive bright fringes. As the phase δ varies across the deformed region the resulting intensity varies, giving information about the value and the variation of the optical thickness across the phasor. If Δ is small, the order of interference N is small, being equal to 0, 1 or 2. Coloured fringes appear when the interferometer is illuminated with white light.

The fundamentals of multiple-beam interferometry and their application to fibres are given in Chapter 4.

Interference microscopes can be divided into two groups: (i) those using reflected light, i.e. dealing with opaque objects such as metals, and (ii) microscopes with transmission systems to deal with transparent objects such as polymer fibres and biological materials. Group (i) gives information about the surface microtopography of the object. Group (ii) gives information about the sample structure and the value of $n_\lambda t$ at every point of the object. If the metric thickness t is measured, the refractive index n_λ is readily calculated. Interference microscopes have high magnification and high resolution only in the up–down (depth) dimension (see Tolansky 1973).

The following two designs are described to show the optical path in two examples of these two groups of microscopes.

(*a*) Interference microscope using reflected light. Tolansky (1944) developed a simple arrangement to be used either as a two-beam or as a multiple-beam system (see Tolansky 1973). In figure 7.2 light from the source A is monochromatized with the filter F and forms an image I at the back focal plane of the microscope objective O. S is the interference system and is composed of an object resting upon an optical flat. This interference system S is illuminated at normal incidence with a parallel beam of monochromatic light. One can see the surface of the object covered with interference fringes and an optical contour map with height changed by $\lambda/2$ (λ is the wavelength of light used) in moving from one fringe to another.

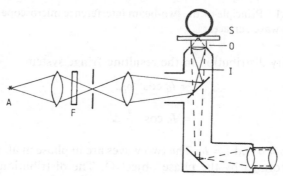

Figure 7.2 An interference microscope using reflected light. (From Tolansky 1973.)

(*b*) Interference microscope using transmitted light. Transparent microscopic objects including polymeric and optical fibres are viewed by transmitted light. Many interference microscopes using transmission systems are available. Tolansky (1973) classified these instruments into (i) those using two microscopes, (ii) those using one modified microscope, (iii) interference objectives and (iv) differential interferometers, and he described representative instruments.

One of the earliest practical interference microscopes in transmission was described by Sirks in 1893. This interference microscope consists of a Jamin interferometer mounted before the optical microscope (see Tolansky 1973). Figure 7.3 shows a schematic diagram of this instrument, A and B are two identical glass blocks, each containing a small area of silver. A parallel beam of light is divided into the two beams 1 and 2 as shown in figure 7.3.

One of these two beams passes through the object O and the other beam passes nearby. Thus a path difference is introduced. After passing through B, light interferes and interference fringes are seen by the microscope. These fringes give information about the object O.

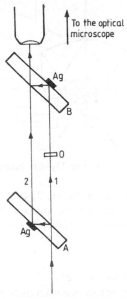

Figure 7.3 Jamin interferometer-type microscope. (From Tolansky 1973.)

For further information about the fundamentals of interference microscopy and their applications to fibre research the following references should be consulted: Tolansky (1948, 1973), Françon (1961), Heyn (1954, 1957), Barer (1955), Stoves (1957), Pluta (1971, 1972, 1982), Fatou (1978), Steel (1986) and Hamza (1986).

In the following section some two-beam interference microscopes and systems are described from the point of view of fibre interferometry.

7.2 Some types of interference microscope

7.2.1 *The Dyson interferometer microscope*

In the Dyson (1950, 1953) microscope, the illuminating cone from a condenser is intercepted by a parallel plate, the surface of which is coated with a thin silver layer (partially silvered). Part of the light is focused on the object. Before passing through the plate, another part is twice reflected by the two surfaces of the plate and, therefore, two beams pass through the object plane. Figure 7.4 shows the optical system of the Dyson (1950) interferometer microscope.

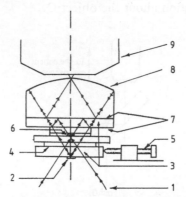

Figure 7.4   Optical system of the Dyson interferometer microscope. (From Dyson 1950.)

1, illuminating cone from condenser; 2, fully silvered spot; 3, nearly plane parallel plates; 4, half-silvered surface; 5, screw movement; 6, object; 7, half-silvered surface; 8, fully silvered surface; 9, microscope objective.

7.2.2   Baker interference microscope

This is a double-refracting interference microscope developed in 1950. Heyn (1957) discussed its use in the study of textile fibres. In the Baker interference microscope a beam of light is separated by a beam splitter consisting of a doubly refractive crystal or plate so that light emerges in two separate rays (the ordinary and extraordinary rays). These rays are polarised in mutually perpendicular planes. The two beams pass through different parts of object space. The refractive indices of fibres can be measured easily with this instrument.

7.2.3   The Interphako interference microscope

This is a suitable instrument for measuring minute path differences in microscopic specimens. Various interferometric methods of observation and measurement can be carried out with this instrument. For examinations in transmitted light† the Interphako consists of a normal-type transmitted light microscope, an intermediate image system and a small Mach–Zehnder interferometer. Figure 7.5 shows the path of the rays in the Interphako (see Beyer and Schöppe 1965). The slit S_p is illuminated by the source of light S. With the aid of the condenser (1) and objective (2), the slit S_p is imaged into the objective's rear focal plane at $S_{p'}$. The intermediate

† See *Description and Instruction Manual* Carl Zeiss Jena, Brochure No 30-G305-2.

image system (3, 4, 5) forms an intermediate image of the object O at O'
and O″ as well as an image of the exit pupil at $S_{p''}$ in the interferometer.
Prism 4 is used to position the image $S_{p''}$ of the pupil at its specific place.
Prism 12 is used for photomicrographic work and the Bertrand lens (13) is
used for observing the pupil.

The interferometer of this instrument consists of the two prisms (6 and 7),
the phaseshifter (8), the rotary wedge (10) exchangeable for the annular
diaphragms, as well as two compensating elements (9 and 11). The two
partial beams produced at the beam-splitting surface S_1 emerge from the
exits and A and B of prism 7, exactly corresponding to each other as regards
their height and direction. The phaseshifter (8) causes the path difference
between $S_1 S_{p_2}$ and $S_1 S_{p_1} S_2$ to be varied by ±15 wavelengths ($\lambda = 500$ nm),
at maximum. Path difference measurements up to thirty orders are possible.
The rotary wedge (10) consists of two glass wedges of identical design,
which may be revolved on the optical axis against each other. Using the
Interphako with appropriate objects a measuring accuracy up to $\lambda/200$
without and up to $\lambda/500$ with a half-shade plate is achieved.

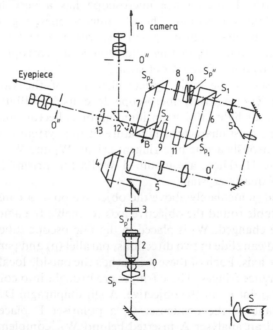

Figure 7.5 The path of the rays in the Interphako interference microscope. (From *Description and Instructions Manual* Carl Zeiss Jena,
Brochure No 30-G305-2.)

S, source of light; S_p, slit; O, slide object; 1, condenser; 2, objective;
3, 4, 5, the intermediate image system; 6, 7, two prisms; 8, phaseshifter;
9, 11, two compensating elements; 10, rotary wedge; 12, prism; 13,
Bertrand lens.

Hamza (1986) and Hamza *et al* (1986) used the Interphako for the measurement of the mean refractive indices, and hence birefringence, of polyester fibres and bicomponent fibres (nylon 6 sheath, nylon 66 core). Both white and monochromatic light were used.

7.2.4   The Pluta polarising interference microscope

Pluta (1965, 1971, 1972) developed a double-refracting interference microscope with a variable amount and direction of wavefront shear and he described the use of this microscope for the study of synthetic polymer fibres. Pluta achieved some improvement in the microinterferometric measurement of refractive indices and double refraction of fibres. This microscope is capable of giving both the uniform and fringe interference fields with a continuously varied amount and direction of lateral image duplication. It can be used for both qualitative observation, using the differential interference contrast (DIC) method, and for quantitative examination (measuring the optical path difference) of various specimens in transmitted light.

The conventional interference microscope has a very limited use in polymer fibre characterisation. The DIC microscopes, e.g. the Nomarski (1955) system, have many advantages in this area. We shall now describe the principle of the double-refracting interference microscope with variable amount and direction of wavefront shear.

Figure 7.6 shows the basic optical elements of the variable double-refracting interference (VDRI) microscope developed by Pluta (1971, 1972). Z is the source of light, Kol is an illuminator collector and D_p is a field diaphragm. The most characteristic feature of this system is a combination of two simultaneously acting birefringent prisms W_1 and W_2 separated by a half-wave plate H. These are modified Wollaston prisms (see Nomarski 1955) made of quartz crystal.

W_1 is located immediately above the objective ob at a constant distance i_1, and is rotatable round the objective axis to enable the amount of image doubling to be changed. W_2 is placed in the microscope tube at a variable distance i_2, and can slide in two directions, parallel (p) and perpendicular (t) to the objective axis. Each of these prisms has the outside localising plane of its own interference fringes. These fringes are brought into coincidence with the back focal point F′ of the objective. A slit diaphragm D located in the front focal plane of the condenser C, a polariser P placed below this diaphragm and an analyser A inserted behind W_2, complement the basic system of the microscope. The slit diaphragm D, the half-wave plate H and the polariser P are rotatable around the objective axis. Figure 7.7 shows the initial orientation of P, A, H and S with respect to the position of W_1 and W_2, E_1 and E_2 denote edges of wedge angles a_1 and a_2 of the upper wedges of the birefringent prisms W_1 and W_2; FF is one of the principal directions of light vibrations of the half-wave plate H and SS is in the direction of the slit S.

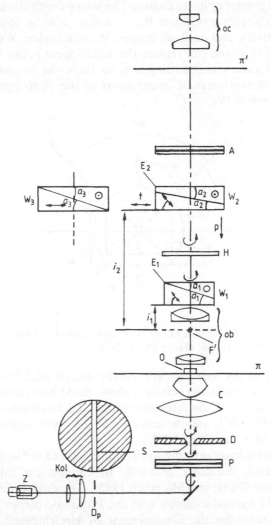

Figure 7.6 The optical system of the double-refracting interference microscope with a continuously variable amount and direction of wavefront shear. (From Pluta 1972.)

In the VDRI microscope, a plane light wave, polarised linearly by the polariser P, leaves the condenser C and is distorted by the transparent object O. A phaseshift corresponding to the optical path difference occurs in the object. The distorted wavefront enters the objective ob, and is split by the prisms W_1 and W_2 into two wavefronts polarised at right angles. When passing through the analyser A, both wavefronts interfere with each other and make visible the transparent object O in a form of two images laterally duplicated to a greater or lesser degree. The image duplication is changed by rotation of the birefringent prism W_1. π and π' are the object and image planes, respectively, and oc is an ocular. W_3 is a typical Wollaston prism with wedge angle a_3 and can replace the birefringent prism W_2.

Pluta (1972) gave vectorial diagrams to show the resultant wavefront shear in the different cases of orientations of the birefringent prisms W_1 relative to the prism W_2.

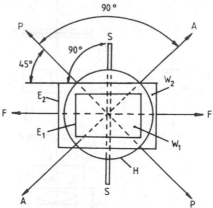

Figure 7.7   Initial orientation of principal elements of the microscope illustrated in figure 7.6. (From Pluta 1972.)

The accuracy of the measurements of the optical path length difference with a fringe field (Wollaston prism) is about 0.05λ and therefore the error in the determination of the refractive index and birefringence cannot be better than 0.003–0.001, and the accuracy of the determination of the fibre diameter is about 1 μm (Pluta 1972).

With the Pluta microscope, monochromatic light ($\lambda = 546$ nm) and white light can be used, the latter primarily to confirm the position of the achromatic fringe (Faust and Marrinan 1955). The differential interference and fringed field method available with the Pluta microscope appears to be particularly suitable for the measurement of birefringence of fibres and films, especially those exhibiting some minor local heterogeneities. The Pluta microscope provides an easy and quick method for measuring the mean refractive indices and the mean birefringence of heterogeneous fibres of regular and irregular transverse sections. It is remarkably useful when measuring fibres of high refractive index values or when using immersion liquid of refractive index differing greatly from that of the fibre (Hamza and Sikorski 1978).

Pluta (1972) discussed the advantages of his microscope from the point of view of fibre microscopy. Extensive optical studies on natural and synthetic fibres using this microscope have been carried out by Hamza and co-workers (see Hamza (1986) and the papers cited therein).

7.2.5 The shearing effect in interference microscopy

The shearing method is based on lateral image splitting. When the splitting is larger than the object, it is called total image splitting. In the event of small lateral image splitting of a magnitude of the order of the resolving limit, it is called differential splitting.

In the Interphako interference microscope, both total and differential image splitting are made possible by using a rotary wedge. A rotary wedge consists of two glass wedges of identical design, which may be revolved about or around the optical axis in opposite directions with reference to each other. The change in the direction caused with the aid of the rotary wedge in only one partial path of the two rays (produced by a beam splitting surface) corresponds to an image splitting in the image plane, which may be set to any value between 0 and $+0.4$ mm in the intermediate image plane in a vertical direction. The shearing method with total image splitting is applied for contrasting and measuring larger objects.

Using the Pluta microscope, a large image shearing in a uniform interference field is possible with the use of high-shearing objectives (polarising interference objectives). The Wollaston prism installed in the interference head of the microscope and that installed in the objective form, together with two crossed or parallel polaroids and a slit, make a kind of double-polarising interferometer (Polskie Zakłady Optyczne (PZO) 1976). The Wollaston prism in the objective is able to rotate about the vertical axis to adjust the magnitude of image shearing. Maximum image shearing (r) is obtained when the angle of refraction of the two prisms has the same orientation. In this case

$$r = r_1 + r_2$$

where r_1 and r_2 are the shearing values produced by the first and second prisms, respectively. If, on the other hand, the second prism is reversed such that its angle of refraction is opposite in direction to that of the first prism then

$$r = r_1 - r_2.$$

It follows that by rotation of the second prism, installed in the objective, around a vertical axis, the image shearing values range from $r_1 + r_2$ to $r_1 - r_2$. The high image shearing effect enables measurement of the optical path difference for both isotropic and anisotropic objects along with their thickness and refractive index.

7.2.6 The Ziess–Linnik interference microscope

An important modification of the Michelson interferometer in the study of the microtopography of surfaces is the Zeiss–Linnik interference microscope. It is a relatively modern interference microscope designed and constructed by Linnik (1933) and shown in figure 7.8. S is the source of illumination, C a condenser lens and G a beam splitter inclined at 45°. Part of the light is reflected towards the objective O_1, illuminates the surface under investigation and reflects back carrying the information to the eyepiece O_2 through G. The light transmitted through G traverses the objective O_1', identical to O_1, reflected by the plane reference mirror M, returns again to O_1, reflected at G, then is directed towards the eyepiece O_2. The two images F_1 and F_2 of the light source S are formed at the focus of the two objectives O_1 and O_1'. The modulated wave carrying information about the irregularities of the surface of the object J and the reference plane wave, interfere in the image P' observed by the eyepiece O_2. A two-beam interferogram is formed, the shape and fringe displacements giving the surface topography.

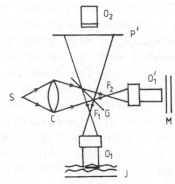

Figure 7.8   The Linnik interference microscope.

Figure 7.9 shows the passage of light in the Zeiss–Linnik interference microscope. Interference takes place between two reflected beams, one from a reference plane mirror and the other from the surface of the object under investigation. The reference mirror is in the form of a cap attached to an objective and at a specific distance. Three caps are provided of reflectivities 30, 60 and 90% for each of the three objectives available, of magnification 10, 25 and 60×. The cap chosen is the closest in reflectivity to that of the sample. The higher the magnification, the nearest will be the cap to the objective in terms of distance. This is also the condition for adjusting the distance between the surface under study and the objective introduced in the passage of light, to illuminate the sample and collect the light reflected from its surface carrying the information.

The range of surface features that could be measured along the direction perpendicular to the object surface is from $\lambda/2$ to 20λ. A thallium spectral line of wavelength $\lambda = 535$ nm is used. A white light source replaces the monochromatic source to form low-order white light fringes, by rotating the reflector I shown in figure 7.9. Precise adjustment takes place by obtaining the monochromatic system of interference fringes, then the white light lamp replaces the thallium lamp and low-order white light fringes are obtained. The thallium lamp then replaces the white light source. By this means, the monochromatic fringes formed are of high contrast and in focus over the whole field of view. A telescope eyepiece for viewing the fringe system formed or a camera is used to record the interferogram. The two-beam arrangement in the Zeiss–Linnik interference microscope has the advantage that both direction and linear dispersion of the two-beam interference pattern can be altered at will, by adjusting a plate introduced in the path of one of the two beams. It is to be noted that one of the main differences between the two-beam interference fringes as formed by the Linnik-type interference microscope and localised Fizeau fringes, is that while the former magnify the surface under investigation first, then interference fringes are formed on it, the latter forms the localised interference pattern close to the interferometer, then a magnified image is recorded on the photographic plate.

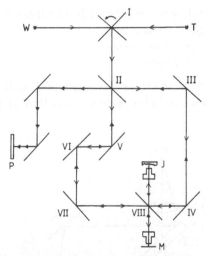

Figure 7.9 Schematic diagram of the passage of light in a Zeiss–Linnik interference microscope.

T, thallium lamp; W, white light; J, object; M, reference mirror; I, opaque mirror (reflector); P, photographic plate; II, half-silvered plate; III, IV, V, VI, VII, opaque reflectors silvered from both sides.

7.2.7 The Mach–Zehnder interferometer

In the Jamin interferometer the front surfaces of the two plates act as beam dividers and the rear surfaces as plane mirrors, but these elements cannot be adjusted independently and the separation of the two beams is limited by the thickness of the plates. In the Mach–Zehnder interferometer (Zehnder 1891, Mach 1892) the beams may be widely separated as the beam dividers and mirrors are separate elements. Figure 7.10 shows the square arrangement and passage of light. S is a source of monochromatic light at the focal plane of a corrected lens L_1, a beam of parallel light is divided at the semi-reflecting surface A_1 of a plane parallel glass plate D_1 into two beams, which after reflection at plane mirrors M_1 and M_2 are recombined at the semi-reflecting surface of A_2 of a second identical plane parallel plate D_2 and emerge to a corrected lens L_2. Let W_1 be a plane wavefront in the beam between M_1 and D_2, W_2 the corresponding plane wavefront in the beam between M_2 and D_2 and W_1' the virtual plane wavefront between M_2 and D_2 which would emerge from D_2 coincident with W_1. At a point P on W_2, the virtual phase difference (Born and Wolf 1980) between the emergent beams is then $\delta = 2\pi nh/\lambda$, where $h = PN$ is the normal distance from P to W_1' and n is the refractive index of the medium between W_2 and W_1'. At the point P' conjugate to P, bright fringes appear for $m\lambda = n\lambda$, $|m| = 0, 1, 2, \ldots$.

Three cases of phase objects arise in practice.

(*a*) Two-dimensional phase objects with no variation of refractive index in the direction of light propagation.

(*b*) Radially symmetric phase objects.

(*c*) Asymmetric phase objects.

For the first case, the phase object has a length L in the direction of light propagation; the refractive index is then a function of z and y only.

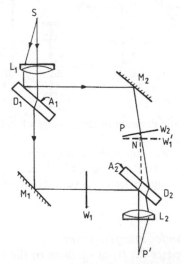

Figure 7.10 The passage of light in the Mach–Zehnder interferometer.

Fringe shift equation

When a phasor is introduced in one beam of a Mach–Zehnder interferometer (figure 7.11) the fringe shift $\delta(y, z)$ is given by

$$\delta(y, z) = \frac{1}{\lambda} \int_{x_0}^{x_1} (n - n_0) \, \mathrm{d}x \qquad n = n(x, y, z)$$

where n_0 is the refractive index in the undisturbed beam and λ is the wavelength. To prove this equation we observe that the optical path of the disturbed ray measured in wavelengths is

$$N_\mathrm{d} = \int_{x_0}^{x_1} \frac{\mathrm{d}x}{\lambda}.$$

The difference $N_\mathrm{d} - N_0$ gives the number of multiples of 2π by which the two rays differ in phase when recombined. This number is just equal to the fringe shift $\delta(y, z)$.

If $n = n(y, z)$ then for a two-dimensional phase object

$$n(y, z) - n_0 = \frac{\delta\lambda}{x_1 - x_0} = \frac{\delta\lambda}{L}.$$

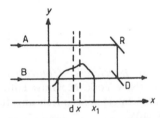

Figure 7.11 The passage of light through the disturbance of a phasor in a Mach–Zehnder interferometer. A, undisturbed ray; R, reflector; B, disturbed ray; D, beam divider.

7.2.8 Leitz interference microscope

This microscope is composed of two separate but identical microscopes: one containing the object under study, J, the other providing an undisturbed wave. The first gives an image which interferes with the reference plane wave provided by the second microscope. Figure 7.12 shows the arrangement and passage of light in the Leitz interference microscope; P_1, P_1', P_2, P_2' are combination prisms that replace the reflectors and beam dividers at the four corners of a Mach–Zehnder interferometer. O_1 and O_1' are two objective lenses corrected for infinity and the correction lens L forms two superimposed images of J and R which are observed with the eyepiece O_2. The plates with parallel faces d_1 and d_2 allow variation of the path difference Δ.

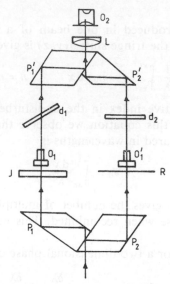

Figure 7.12   Arrangement and passage of light in a Leitz interference microscope.

References

Barer R 1955 *Phase Contrast, Interference Contrast and Polarizing Microscopy* Analytical Cytology Series (New York: McGraw-Hill)

Beyer H and Schöppe G 1965 Interferenzeinrichtung für durchlicht Mikroskopie *Jenaer Rundschau* **10** 99–105

Born M and Wolf E 1980 *Principles of Optics* 6th edn (London: Pergamon) p315

Dyson J 1950 *Proc. R. Soc.* A **204** 170

—— 1953 *Nature* **171** 743

Fatou J E 1978 Optical microscopy of fibres in *Applied Fibre Science* ed. F Happey vol. 1 (London: Academic) Ch. 3

Faust R C and Marrinan H J 1955 *Br. J. Appl. Phys.* **6** 351

Françon M 1961 *Progress in Microscopy* (London: Pergamon) pp94–128

Hamza A A 1986 *J. Microsc.* **142** 35

Hamza A A, Fouda I M and El-Farahaty K A 1986 *Int. J. Polym. Mater.* **11** 169

Hamza A A and Sikorski J 1978 *J. Microsc.* **113** 15

Heyn A N J 1954 *Fibre Microscopy* (New York: Interscience)

—— 1957 *Textile Res. J.* **27** 449

Linnik W 1933 *C. R. Acad. Sci. URSS* **1** 21

Mach L 1892 *Z. Instrkde.* **12** 89

Nomarski G 1955 *J. Phys. Radium, Paris* **16** 95

Pluta M 1965 *Przeglad Włókienniczy* **19** 261

—— 1971 *Opt. Acta* **18** 661

—— 1972 *J. Microsc.* **96** 309

Pluta M 1982 *Mikroskopia Optyczna* (Warszawa: Państwowe Wydawnictwo Naukowe) (in Polish)
Polskie Zakłady Optyczne (PZO) Instruments 1976 *Biolar Polarizing Interference Microscope, Description and Instruction Manual*
Steel W H 1986 *Interferometry* (Cambridge: Cambridge University Press)
Stoves J L 1957 *Fibre Microscopy* (London: National Trade Press)
Tolansky S 1948 *Multiple-Beam Interferometry* (Oxford: Clarendon)
—— 1973 *An Introduction to Interferometry* (London: Longman) pp210–23
Zehnder L 1891 *Z. Instrkde.* **11** 275

8 Back-scattering of Light Waves from Fibres

8.1 The case of a beam of light incident perpendicular to the fibre axis

Optical fibres currently being considered for use as the transmission medium in optical communication systems consist of two concentric dielectric cylinders of very great length (kilometres) and very small external diameter $\approx 125\ \mu$m. The inner cylinder, the core, is composed of a material of refractive index n_{core} slightly more than that of the outer cylinder n_{clad} of the cladding. The refractive indices of core and cladding and the diameter of the core are the three basic parameters that specify the transmission characteristics of the fibre. A simple non-destructive technique to measure the refractivity and diameter which could be used as an on-line system for monitoring and regulating these properties as fibres are drawn during their production would be valuable. As a first step towards this goal, we describe the determination of the refractive index and diameter of uncladded optical fibres following a method suggested by Presby (1974). The technique is based on an analysis of back-scattered light when a beam from a CW laser impinges upon the fibre, being incident perpendicular to the fibre axis. A geometrical-optics analysis shows that the position of a sharp cut-off in the radiation pattern determines the refractive index of the uncladded fibre, whereas the distance between certain successive minima gives the diameter.

We present here the result of applying the findings of Presby to uncladded fibres of soda glass, silica and Pyrex of diameters in the range of $100-300\ \mu$m. As in Presby's work only one single internal reflection has been considered.

8.1.1 Back-scattering analysis
Let a beam of collimated, monochromatic light of wavelength λ be incident upon an uncladded specimen of glass fibre. Light polarised parallel to the axis of the fibre must be used for these measurements, because calculations

130

of the combined Fresnel coefficients show that light polarised perpendicular to the axis of the fibre has minimum irradiance when it emerges from the fibre at angles very close to ϕ_m as shown in figure 8.1.

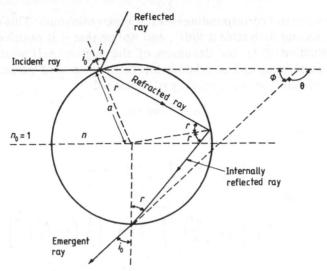

Figure 8.1 Incident, reflected and refracted rays.

In the geometrical optics approximation, when a ray hits the fibre it gives rise to a reflected and a refracted ray as shown in figure 8.1. In an isotropic, homogeneous medium, these rays are straight lines, and at the boundaries between two isotropic media the directions of the lines are altered, as predicted by the law of reflection and Snell's law for refraction. Use of these laws permits the rays to be traced through the fibre, so that the emergent rays that contribute to the scattering may be found. In addition, account must be taken of interference effects that, through Fresnel coefficients, determine the irradiance in the pattern.

If i is the angle of incidence of a ray at a given point and r is the angle of refraction, the ray is rotated through an angle $i - r$ upon entering the fibre. On reflection at the back surface, it suffers a further rotation of $\pi - 2r$ and on emergence another rotation through an angle of $i - r$. Thus the total deviation θ of the ray is $\pi + 2i - 4r$. To calculate the minimum angular deviation we deduce $d\theta/di$ and set it equal to zero. The result, in which we let $n_0 = 1$ and n be the refractive index of the fibre, gives the angle of incidence corresponding to the minimum deviation

$$\theta = \pi + 2i - 4 \sin^{-1}\left(\frac{\sin i}{n}\right) \tag{8.1}$$

$$\frac{d\theta}{di} = 2 - \frac{4 \cos i}{(n^2 - \sin^2 i)^{1/2}} \tag{8.2}$$

when $d\theta/di \to 0$, hence $i = i_m$, and

$$\cos i_m = \left(\frac{n^2 - 1}{3}\right)^{1/2}.$$ (8.3)

This is the value of i corresponding to a stationary minimum. This is seen by taking the second derivative $d^2\theta/di^2$, and noting that it is positive.

From equation (8.3), the definition of the angular half width of the scattering pattern ϕ_m and its relationship to the measured parameters L_m and h, we obtain from the geometry of figure 8.2

$$L_m = h \tan \phi_m$$ (8.4)

since $\phi = \pi - \theta$

$$\phi_m = 4 \sin^{-1}\left(\frac{\sin i_m}{n}\right) - 2i_m$$ (8.5)

and finally we get

$$\phi_m = 4 \sin^{-1}\left[\frac{2}{n\sqrt{3}}\left(1 - \frac{n^2}{4}\right)^{1/2}\right] - 2 \sin^{-1}\left[\frac{2}{\sqrt{3}}\left(1 - \frac{n^2}{4}\right)^{1/2}\right].$$ (8.6)

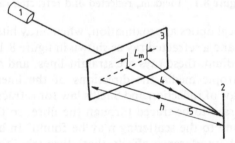

Figure 8.2 Set-up to observe back-scattered light.
1, CW He–Ne laser beam at the wavelength 632.8 nm; 2, uncladded optical fibre; 3, viewing screen; 4, incident radiation; 5, back-scattered radiation.

Equation (8.6) indicates that the sharp cut-off in the pattern of back-scattering of laser radiation when it interacts with the glass fibre allows the calculation of the refractive index of the fibre.

8.1.2 Determination of the refractive index of the fibre material; experimental procedure

The set-up used to observe the back-scattering pattern is shown schematically in figure 8.2. Light from a CW He–Ne laser passes through an aperture in an opaque viewing screen and is incident upon the fibre. The back-scattered light falls on the same screen and is photographed. Light scattered in other directions is blocked from entering the camera. The complete back-scattered light is localised in a range of angular deviation of $\phi_m \simeq \pm 20°$ from the incident direction. The complete pattern is symmetrical about $\phi = 0$, λ being 6328 Å. Figure 8.3 shows back-scattering patterns of soda glass fibre of diameter about 200 μm.

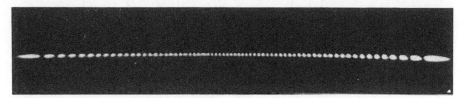

Figure 8.3 The back-scattered pattern of a soda glass fibre of diameter about 200 μm.

8.1.3 Back-scattering of laser radiation from a cladded fibre

For low-loss optical fibres used as the transmission medium in optical communication systems it is possible to determine the physical properties of such cladded fibres with the help of back-scattering of light incident perpendicular to the fibre axis.

Ho *et al* (1975) dealt with the problem of back-scattering in the case of optical fibres where $n_{core} > n_{clad}$ which is the necessary condition for total internal reflection within the core at the interface with the cladding.

Back-scattering from cladded fibres is characterised by two sharp cut-offs.

The case of an optical fibre with $n_{clad} > n_{clad}$, considering single reflection

The pattern resulting from back-scattering can be studied following a geometrical-optics approximation. The trajectory of incident rays is followed through refractions and reflections as shown in figure 8.4 where the relevant angles are defined. These are related by

$$\theta = \pi - 4\gamma' + 2i + 2i' - 2\gamma = \pi - \phi.$$

By setting

$$\left.\frac{d\phi}{di}\right|_{i=i_m} = 0$$

one can show that i_m satisfies

$$\frac{2\cos i_m}{[(n_2' r_2/r_1)^2 - \sin^2 i_m]^{1/2}} + \frac{\cos i_m}{(n_1'^2 - \sin^2 i_m)^{1/2}} - \frac{\cos i_m}{[(n_1' r_2/r_1)^2 - \sin^2 i_m]^{1/2}} = 1 \tag{8.7}$$

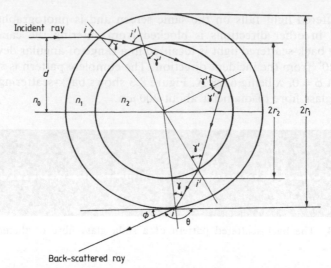

Figure 8.4 Geometry for a back-scattered ray considering a single reflection inside the core for a fibre of optical parameters: n_1, refractive index of the cladding; n_2, refractive index of the core; r_1, radius of the whole fibre; r_2, radius of the core.

where $n_2' = n_2/n_0 = n_{\text{core}}$ and $n_1' = n_1/n_0 = n_{\text{clad}}$, n_0 being the index of the medium surrounding the fibre and being equal to one.

Equation (8.7) can hardly, in general, be solved analytically for i_{m}, but a close approximation can be obtained for $r_1 \approx r_2$. In this case one obtains for the maximum back-scattering angle

$$\phi_{m=1} = 4 \cos^{-1}\left\{\frac{1}{n_c}\left[\frac{4}{3}\,n_c^2 - \frac{4}{3}\left(\frac{r_1}{r_2}\right)^2\right]^{1/2} - 2\cos\left\{\left[\frac{1}{\sqrt{3}}\left(n_i\,\frac{r_2}{r_1}\right)^2 - 1\right]^{1/2}\right\}\right.$$

$$- 2\cos^{-1}\left\{\frac{1}{n_s}\left[n_s^2 + \frac{1}{4}\,n_c^2 - \frac{4}{3}\left(\frac{r_1}{r_2}\right)^2\right]^{1/2}\right.$$

$$\left.\left. + 2\cos^{-1}\left[\left[\frac{1}{n_s}\left\{n_s^2 + \frac{4}{3}\left[\frac{1}{4}\,n_c^2\left(\frac{r_2}{r_1}\right)^2 - 1\right]\right\}\right]^{1/2}\right]\right]\right\}$$

where $n_c = n_{\text{core}}$ and $n_s = n_{\text{clad}}$.

For an unclad fibre where $r_1 = r_2$, $n_c = n_s = n$ and $n_0 = 1$, the last formula leads to the formula derived previously due to Presby (1974).

The case of optical fibres, considering two internal reflections
The following presents the analysis of back-scattering considering two internal reflections occurring within the fibre core. The trajectory of the

incident rays is followed through refractions and reflections as shown in figure 8.5 where the relevant angles are defined. These are related by

$$\theta = 2\pi - 6\gamma' + 2i + 2i' - 2\gamma = 2\pi - \phi.$$

By setting

$$\left.\frac{d\phi}{di}\right|_{i=i_m} = 0$$

one can show that i_m satisfies

$$\frac{3\cos i_m}{[(n_c r_2/r_1)^2 - \sin^2 i_m]^{1/2}} + \frac{\cos i_m}{(n_s^2 - \sin^2 i_m)^{1/2}} - \frac{\cos i_m}{[(n_s r_2/r_1)^2 - \sin^2 i_m]^{1/2}} = 1.$$

Assuming $r_1 \approx r_2$, the maximum back-scattering angle ϕ_m is given by

$$\phi_{m=2} = 6\cos^{-1}\left\{\frac{1}{n_c}\left[\frac{9}{8}n_c^2 - \frac{9}{8}\left(\frac{r_1}{r_2}\right)^2\right]\right\}^{1/2} - 2\cos^{-1}\left\{\frac{1}{2\sqrt{2}}\left[\left(n_c\frac{r_2}{r_1}\right)^2 - 1\right]^{1/2}\right\}$$

$$- 2\cos^{-1}\left\{\frac{1}{n_s}\left[n_s^2 + \frac{1}{8}n_c^2 - \frac{9}{8}\left(\frac{r_1}{r_2}\right)^2\right]^{1/2}\right\}$$

$$+ 2\cos^{-1}\left\{\frac{1}{n_s}\left[n_s^2 + \frac{1}{8}\left(n_c\frac{r_1}{r_2}\right)^2\frac{1}{9} - \frac{9}{8}\right]^{1/2}\right\}.$$

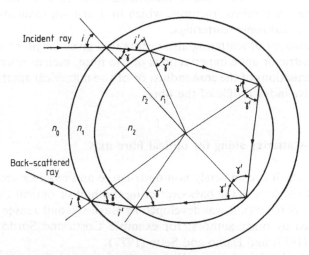

Figure 8.5 Geometry for a back-scattered ray considering two reflections inside the core for a fibre of optical parameters: n_1, refractive index of the skin; n_2, refractive index of the core; r_1, radius of the core; r_2, radius of the whole fibre.

In general for multiple reflections m inside the core, we have obtained the maximum back-scattering angle ϕ_m from the following equation:

$$\phi_m = 2(m + 1)\cos^{-1}\left[\left[\frac{1}{n_c}\left\{\frac{(m+1)^2}{m(m+2)}\left[n_c^2 - \left(\frac{r_1}{r_2}\right)^2\right]\right\}^{1/2}\right]\right]$$

$$- 2\cos^{-1}\left[\left(\frac{(n_c r_2/r_1)^2 - 1}{m(m+2)}\right)^{1/2}\right]$$

$$- 2\cos^{-1}\left[\left[\frac{1}{n_s}\left\{n_s^2 + \frac{(m+1)^2}{m(m+2)}\left[\frac{n_c^2}{(m+1)^2} - \left(\frac{r_1}{r_2}\right)^2\right]\right\}^{1/2}\right]\right]$$

$$+ 2\cos^{-1}\left\{\frac{1}{n_s}\left[n_s^2 + \frac{(m+1)^2}{m(m+2)}\left(\frac{(n_c r_2/r_1)^2}{(m+1)^2} - 1\right)\right]^{1/2}\right\}.$$

We are now able to calculate n_c from the knowledge of $\phi_m = \tan^{-1}(L_m/h)$, $2L_m$ being the actual length between the points giving sharp cut-off and h the distance between the fibre and the centre of the pattern; n_s is determined by the Becke-line method (see Hartshorne and Stuart 1970). Determination of ϕ_m for $m = 1$ and $m = 2$ enables calculation of n_c and n_s for known values of r_1/r_2. A computer program performs the calculation.

8.1.4 Light scattering of a beam incident through the fibre along its axis

Light scattering in optical fibres is due to Rayleigh scattering (the ultimate cause of loss in the fibre) and to inhomogeneities such as micro-bubbles, micro-crystals and micro-fractures, which in a first approximation may be considered as isotropic scatterings.

A large amount of scattering also occurs in a given length of fibre due to small geometric or optic imperfections of the fibre, such as micro-bending, smooth fluctuations of the core radius, or of the numerical aperture Δ or of the refractive index profile of the fibre.

8.2 Back-scattering along the optical fibre axis

A method which is completely non-destructive and requires access only to one end of the fibre is the back-scattering method or optical time domain reflectometery (OTDR). It was developed by Barnoski and Jensen (1976) and implemented by other authors, for example Costa and Sordo (1977a,b), Personick (1977) and Daino and Sette (1977).

This method is based on sending a pulse into a fibre: a fraction of the power scattered by the fibre is guided backwards and the sequence of echo pulses forms an envelope pulse which is received and analysed at the same input end.

8.2.1 Theory

Light travelling in a fibre undergoes a distributed scattering, i.e. approximately isotropic Rayleigh scattering. Considering only this scattering the amount of power scattered $P_s(z)$ at a distance z from the input end in a section of length dz is

$$P_s(z) = \gamma_s P(z)\, dz$$

where γ_s is the Rayleigh scattering loss coefficient in m^{-1}; it is assumed constant although in general it may be position dependent due to inhomogeneities in the composition of the fibre material.

Now for a given wavelength λ, the light intensity at a distance z along the fibre is given by

$$P(z, \lambda) = P(0, \lambda)\exp\left(-\int_0^z \gamma(\lambda, z)\, dz\right)$$

where $P(0, \lambda)$ is the optical power launched into the waveguide and $\gamma(\lambda, z)$ is the attenuation coefficient per unit length which may be in general position dependent. An average attenuation coefficient may be defined as

$$\bar{\gamma}(\lambda) = \frac{1}{z}\int_0^z \gamma(\lambda, z)\, dz.$$

Accordingly

$$P(z, \lambda) = P(0, \lambda)\exp(-z\bar{\gamma}(\lambda)).$$

Assuming an approximately isotropic angular distribution for the scattered power, the fraction of captured power S is given by the ratio of the solid angle of acceptance of the fibre to the total solid angle (this holds for step-index fibres but is only approximately true for graded-index fibres)

$$S = \frac{\pi\Delta^2}{4\pi n_0^2} = \frac{\Delta^2}{4n_0^2}$$

where Δ is the value of the fibre numerical aperture, $\Delta = (n_0^2 - n_1^2)^{1/2}$, n_0 is the core refractive index and n_1 is that of the cladding.

Therefore the power back-scattered between z and $z + dz$ is

$$P_{bs}(z) = \gamma_s S P(z)\, dz.$$

It is a fraction of the total scattered light at distance z from the input end in a section of length dz. Its direction is backward towards the entrance end of the fibre and governed by the acceptance angle. As it moves back, it traverses the fibre and also suffers from attenuation.

The power scattered between z and $z + dz$ reaching the detector, assuming a coupling efficiency η will be given by

$$P_{bsd}(z) = \eta P_{bs}(z)\exp\left(-\int_0^z \gamma'(z)\, dz\right)$$

where γ' is the attenuation coefficient for the backward propagating light.

Substituting for

$$P_{bs}(z) = \gamma_s SP(z) \, dz$$

$$= \gamma_s SP(0)\exp\left(-\int_0^z \gamma(z) \, dz\right)$$

we have

$$P_{bsd}(z) = \eta\gamma_s SP(0)\exp\left(-\int_0^z (\gamma(z) + \gamma'(z)) \, dz\right) dz.$$

The two coefficients of forward and backward attenuation could be considered to be equal:

$$P_{bsd}(z) = \eta\gamma_s SP(0)\exp(-2\bar\gamma z) \, dz.$$

The power generated at z is detected after time $t = 2z/v_g$ where v_g is the group velocity of light in the fibre. If the probe pulse has a width ΔT, the total power $P(t)$ falling on the detector at the time t is obtained by summing the last equation in an interval $\Delta z = v_g \Delta T/2$.

Considering $\exp(-\bar\gamma z)$ constant over this length and substituting for $z = v_g T/2$

$$P(t) = \int_0^{v_g T/2} P_{bsd}(z) \, dz$$

$$= \eta\gamma_s SP(0)\exp(-2\bar\gamma v_g t/2)(v_g \Delta T/2)$$

$$= \eta\gamma_s \frac{c}{2n} \Delta TSP(0)\exp(-2\bar\gamma v_g t/2)$$

provided $P(0)$ is constant during ΔT, otherwise its average value must be considered. Therefore the return waveform has an exponential shape from which the total loss coefficient may be evaluated

$$\frac{P(t_1)}{P(t_2)} = \exp\left(-\bar\gamma \frac{c}{n}(t_2 - t_1)\right) \to \bar\gamma = -\frac{n[\ln P(t_1) - \ln P(t_2)]}{c(t_2 - t_1)}.$$

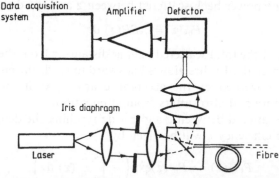

Figure 8.6 The experimental arrangement of the back-scattering measurement apparatus. (After Costa and Sordo 1977b.)

An experimental arrangement used by Costa and Sordo (1977b) for back-scattering measurement is as shown in figure 8.6.

Costa and Sordo (1977b) used a special cell shown in figure 8.7 to reduce reflection from the fibre input face. It is filled with index matching liquid ($n_L = n_{core}$) and includes a beam splitter. The fibre is inserted into the cell through a small hole by means of micromanipulators.

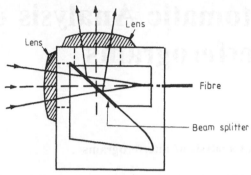

Figure 8.7 An index matching cell used for the reduction of the reflection from the fibre input. (After Costa and Sordo 1977b.)

References

Barnoski M K and Jensen S M 1976 *Appl. Opt.* **15** 2112
Costa B and Sordo B 1977a *CSELT Rep. Tec.* **5** 75
—— 1977b *Third European Conf. on Optical Communication, Munich, September 1977*
Daino B and Sette D 1977 *Eurocon, Venice, May 1977*
Hartshorne N H and Stuart A 1970 *Crystals and the Polarising Microscope* (London: Edward Arnold) pp559–63
Ho P S, Mahric M E and Epstein M 1975 *Appl. Opt.* **14** 2598
Personick S D 1977 *Bell Syst. Tech. J.* **50** 355
Presby H M 1974 *J. Opt. Soc. Am.* **64** 280

9 Automatic Analysis of Interferograms

9.1 The steps of analysis of interferograms

Quantitative analysis of images seen in optical microscopes is limited by the microscopist's eye and brain. The application of digital electronics to microscope-image analysis promises future development of a class quite different. Rosen (1984) reviewed the procedures for insertion of electronics into optical microscopy. His review dealt with objects examined by optical microscopy leading to picture acquisition, digitisation and storage of images, followed by picture analysis, recording and reporting.

This chapter is concerned with insertion of electronics into interference microscopy. Our object is an interferogram, whether two-beam or multiple-beam, single-pass or double-pass interference pattern. Automatic analysis of such interferograms is explained and the deduction of the refractive index profile of fibres is accomplished. Previous investigators who applied automatic analysis of interferograms for deduction of the refractive indices of fibres are mentioned here with a brief account of their contributions. A detailed account is given after reviewing the sequence of procedures given by Rosen (1984), as applied to interference microscopy.

Wonsiewicz *et al* (1976) developed a machine-aided technique of data reduction for interference micrographs, applying the interferometric slab method. Their technique consists of digitising the interferogram with a scanning microdensitometer followed by computer determination of the position of the centre line of each fringe. The data obtained are then converted to refractive index and fibre radius, which is the desired index profile. They also developed the necessary computer program to fit the data to the appropriate power law functions. Presby *et al* (1978) used an automated set-up of a video camera, a digitiser and desk computer to process the output of the interference microscope using the interferometric slab method. They managed to deduce the index profile of graded-index fibres. Boggs *et al* (1979), Presby *et al* (1979) and Marcuse and Presby (1980) used the transverse interferometric method, the fibre being immersed in a matching liquid and illuminated perpendicular to its axis, and described an experimental arrangement for automatic single-pass profile measurements. A Leitz dual-beam, single-pass, transmission interference microscope is used with a video camera and video analysis system. Their measurement procedure involves video detection and digitisation of interference fringes under computer control.

Following Rosen's (1984) review, the sequence of procedures in automatic analysis of interferograms in interference mocroscopy is as follows.

(*a*) Construction of the interferometer and placing it on the microscope stage.

(*b*) Acquisition of a magnified image of the interferogram. The image may be formed and seen visually or on a photographic film or on a television screen. It must persist long enough for quantitative data to be extracted.

(*c*) Extraction of numerical information from the image and placing it in an interim store. Measurements are made with an eyepiece graticule or on a permanently recorded photograph and are transmitted to a calculator or computer.

(*d*) Analysis of the data resulting from the measurements.

(*e*) Recording or reporting the results of the analysis in a permanent form.

9.1. Picture acquisition
The microscopist, whether looking down a microscope or at a photograph or television monitor screen, acquires a picture of the whole field of view at once. The microscopist's retinal cells respond together, in parallel, to the illuminated area of vision. A photographic camera images the scene onto the film by parallel acquisition. In a standard television camera, the picture is converted into an electrical signal by a procedure which reads it serially, as a point progressing along a line, with successive lines covering the picture. Recently introduced alternatives are instruments using diode array detectors or charge coupled devices arranged in an array.

9.1.2 Digitisation and storage of the image
Although a picture can be electronically stored in analogue form, for example on a video tape, we are concerned here only with systems providing image analysis, and in these it is necessary to digitise the picture prior to analysing it. There are two aspects of digitisation, (i) with respect to position and (ii) with respect to light value. This means that the interferogram which is formed in the image plane of the interference microscope or

the interference set-up in transmission or at reflection, is to be replaced by a set of picture elements, termed pixels, each of which can be specified by two spatial coordinates and by a light-intensity value for monochromatic interference systems. It is supplemented by a chrominance index for white light fringes. The fineness of the digitisation depends on the number of pixels per unit area and the number of distinguishable levels of light intensity.

Spatial digitisation

The pixels are arranged as a grid, the most common being the rectangular grid. The optical resolution which can be attained is related to the density of pixels in the grid. A value of 500×500 for a square grid may be the useful limit.

The following is a derivation of an expression for the number of pixels P in a line of length l on the image in terms of the sampling frequency f_s. A point in an illuminated object onto which a microscope is focused has a corresponding diffraction pattern in the image plane. For a lens with a circular aperture, the diffraction pattern is the Airy disc in which the intensity varies according to the function $(2J_1(x)/x)^2$ where $J_1(x)$ is the first-order Bessel function of x which is the distance from the centre of the pattern. According to Hopkins (1943), the Airy formula can be approximated to a Gaussian function, $\exp(-x^2/2r^2)$, which would also accommodate second-order effects when high numerical apertures are used. In the Gaussian function, r is the radius at which the intensity in the imaged diffraction pattern has fallen to 60% of its value at the centre. The spectrum of the spatial frequency of the imaged pattern is given by the Fourier transform of the Gaussian function (see Eccles *et al* 1976a). Clearly a line in the image is produced by convolving the pattern of illumination which constitutes a line across the object with the diffraction pattern of a point source. According to Rosen's treatment, the number of pixels P in a line of length l on the image is given by

$$P = 3l/\pi m(0.22\lambda/\text{NA}) \tag{9.1}$$

where NA is the numerical aperture of the microscope objective, m is the magnification and λ is the wavelength of the monochromatic light used. Eccles *et al* (1976a,b) described a programmable flying-spot microscope that can produce numerical data by digitising the field of view. They used the formula

$$P = 3l/\pi[r_1^2 + (0.22\lambda m/\text{NA})^2]^{1/2} \tag{9.2}$$

where r_1 is the radius of the scan tube at which the intensity $I = 0.6I_{\text{centre}}$.

For a microscope objective with $\lambda = 5000$ Å, with numerical aperture of 1.25 and magnification of 1200, using a high-power oil immersion objective and forming an image 5 cm wide in the television camera, the appropriate number of pixels per line is $P = 452$ (equation (9.1)). This calculation takes no account of the spatial filtering introduced by the electron beam reading the latent image within the camera. A pixel density of 256×256 is probably adequate to capture all the information available. In practice the x and y coordinates of each pixel will be represented in the automated system by a string of bits (zeros and ones) with the restriction that an 8-bit string provides 256, i.e. 2^8 values, a 9-bit string only 512 values. Many standard components are constructed to process with highest efficiency strings of predetermined length 8 bits, 12 bits, ... For automated microscope systems there was a tendency in 1970 to work with 256×256 grids of pixels; now grids containing between 2^9 and 2^{12} rows and columns are available.

Signal digitisation
In any practical situation there is an upper limit to the number of levels into which it is worthwhile digitising the signal. Using binary arithmetic, the signal may be described by a string of m bits so that it would be divisible into N levels where $2^m = N$. If the ratio of signal to noise in the detector is s/n, then according to Billingsley (1971) $N = (s/n)/1.4$. A high-quality electronic apparatus will work with a value of 100 for s/n, $N = 71$ levels. In binary expression, this number requires 7 bits, capable of expressing 128 levels. A television microscope working with $s/n = 300 : 1$, with $N = 214$, would justify an 8-bit coding.

Storage of the image
Formerly, pictures could be stored only on photographs or on video tapes. Advances in large-scale integrated circuitry, as well as the availability of storage discs of all sizes, make it feasible to store large libraries of pictures in digital form. If a picture is digitised as a grid of 512×512 pixels, and the light signal at each pixel is coded as an 8-bit word, the capacity required to store the complete picture is 0.25 Mbyte $\approx 2.1 \times 10^6$ bits. This is a large amount of information—equivalent to a short-to-medium length book—but storage discs, even small floppy discs, can easily accommodate this. Of more immediate importance for automated interference micro-scopy is the possibility of holding such a picture in an interim store from which it can be extracted and analysed. This means holding it in a random access memory (RAM). Now it is quite practicable to work with RAMs with a capacity of several megabits.

9.1.3   Picture analysis: automatic analysis of interferograms and deduction of the index profile of fibre using the interferometric slab method

In Chapter 3, the theory of the interferometric slab method was presented. An expression can be derived relating the refractive index of the fibre core

$n(x, y)$ to the fringe shift $S(x, y)$ as follows:

$$n(x, y) = n_{\text{clad}} + \frac{\lambda S(x, y)}{Dt} \tag{9.3}$$

D being the interfringe spacing and t the slab thickness. The fringe shift can be measured with a reticle in the eyepiece of the microscope and the index computed from equation (9.3), or the fringe shift can be measured on a photograph of the microscopic image, the interferogram being as reported by Wonsiewicz *et al* (1976) and Presby *et al* (1978). The Wonsiewicz method is based on determining the set of Cartesian coordinates that describes the shape of the fringes.

Automatic location of the fringes was performed in the following way, schematically illustrated in figure 9.1.

(*a*) The negative was digitised and encoded on non-magnetic tape by a facsimile device.

(*b*) The magnetic tape was read by a general-purpose computer and the positions of the fringes determined.

(*c*) The *x, y* coordinates of the fringes were written on a disc memory file for use in subsequent computations.

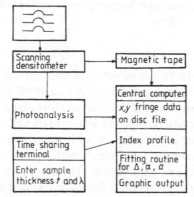

Figure 9.1 Schematic diagram of the automated analysis technique.

For encoding the interferogram a high-quality scanning microdensitometer was used, of resolution 40 lines/mm and subtle gray scale of 256 level. The image was encoded by fastening the transparency to the surface of a drum, which was rotated so that a stationary beam of light scanned across it. The intensity of the transmitted light was recorded using photodetectors, as a series of *n* integers ranging from 0 to 255 which were proportional to the optical density of each *n* point scanned on the film. The scan was repeated line by line to cover the desired portion of the interferogram. The interferogram was scanned at the lowest available resolution of 10 lines/mm with 220 points recorded per line and 250 lines were recorded per interferogram. The process of finding the fringes was achieved by scanning the negative perpendicular to the fringes. A line encoded by the facsimile device is displayed in figure 9.2 as an optical density versus position profile. The fringes are located on a given scan line.

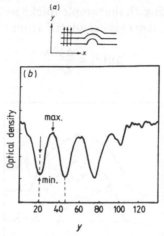

Figure 9.2 (*a*) Scanning the fringes in the direction shown results in the optical density versus *y* curve in (*b*).

The output of the fringe location program is the array of *x*, *y* coordinates of the centre of the scanned fringes. A plot of such an array is shown in figure 9.3. This data array is stored on a disc file and forms the input to the program that calculates the index profile.

9.2 Calculation of the index profile

An interactive program is used to calculate the index profile $n(r)$ from the *x*, *y* fringe data, $n(r)$ is the index value at a distance *r* from the core centre minus the cladding index. Three steps are followed.

(*a*) Baseline. To calculate Δn, the *y* value corresponding to n_{clad} for each of the fringes selected is determined. The *y* value for the cladding of the central fringe is used as a baseline for determining the fringe displacement Δy as shown in figure 9.3 and the cladding index *y* value for each of the selected fringes is used to determine the reference arm wavefront slope of the interferometer. This slope determines the fringe spacing *D* in regions where the index is uniform.

(*b*) Locate fibre axis (the location of the centre of the core (x_0, y_0)). The point x_0 may be defined as the mid-point of the *x* coordinates at which

$y = \frac{1}{2} y_{max}$. Having determined (x_0, y_0), the radial distance r from the fibre axis at each point (x, y) on the central fringe can be calculated from $r^2 = (x - x_0)^2 + (y - y_0)^2$. r can be expressed in micrometres taking into account the microscope's magnification.

(c) Determine the index profile. For each point (x, y) on the central fringe, the fringe displacement Δy and radius r are calculated. Together with the interfringe spacing D, the sample thickness t and the wavelength of monochromatic light λ, $\Delta n(r)$ is calculated from

$$\Delta n(r) = \frac{\Delta y}{D} \frac{\lambda}{t}. \tag{9.4}$$

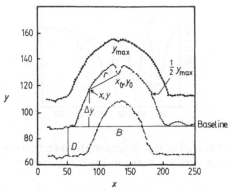

Figure 9.3 The (x, y) array of the position of the centre line of each fringe determined from the photoanalysis programs.

Figure 9.4 is the refractive index profile. The full curve is the least squares fit of the equation

$$\Delta n(r) = \begin{cases} \Delta n_0 \left[1 - \left(\dfrac{r}{a} \right)^{\alpha} \right] & r < a \\ 0 & r \geqslant a. \end{cases}$$

The index depression at the centre is due to loss of GeO_2 from the inside layers of the $GeO_2–SiO_2$ core material during modified chemical vapour deposition, the process of fibre fabrication. This is termed the central dip.

Presby *et al* (1978) used a video camera, a digitiser and desk computer, shown in figure 9.5, to process the output of the interference microscope directly, applying the interferometric slab method. The video camera looks into the interference microscope, its electrical output signal is sent to a digitiser, which functions as an analog-to-digital converter of 8-bit accuracy after addressing specific, preselected points in the video field, the selection being controlled by the computer. Data acquisition proceeds as follows. The computer directs the digitiser to collect light-intensity information on

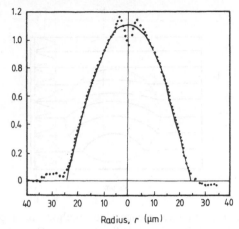

Figure 9.4 The refractive index profile determined from an interferogram by automatic analysis. The full curve is the least squares fit to the data. $\alpha = 1.97$, $\Delta = 0.0076$, $r_{core} = 24.0\ \mu m$.

successive vidicon scan lines such that the sample points fall on a vertical line. The digitiser samples the light-intensity information along a vertical line near the centre of the core. The wavy line shown in figure 9.6 is the light intensity sampled along the central sample line. The computer determines the fringe positions by searching for the minimum light level whose location it pin-points by least square fitting of a parabola using a number of points in the vicinity of the minimum. The computer directs the vertical sample line to collect information on either side of the core, which is then used to determine the interfringe spacing. Then it advances the sample line in small increments, measuring the displacement of the fringe which is recorded as a function of the radial coordinates r measured from the core centre. The resulting function $S(H)$ equivalent to Δy is used to compute $n(r) - n_{clad}$ according to equation (9.4). The index distribution is finally sent to the plotter.

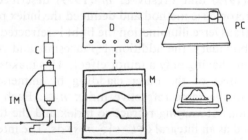

Figure 9.5 Set-up to perform automatic index profiling with a single-pass interference microscope: IM, interference microscope; C, vidicon camera; D, video digitiser; M, display monitor; PC, programmable calculator; P, plotter.

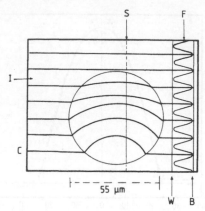

Figure 9.6 Two-beam interference slab method for a graded-index fibre and encoded display of sample line S, cursor C and intensity wave form F. The cladding–matching-oil interface is at I (white reference W and black reference B).

9.3 Automatic analysis of interferograms and deduction of the index profile of fibres using a transverse interference pattern

In Chapter 3 an expression relating the fringe shift $S(y)$ to $n_m(r)$ was derived in the form

$$S(y) = \frac{2D}{\lambda} \int_y^R \frac{\Delta n_m(r)r \, dr}{(r^2 - y^2)^{1/2}} \qquad (9.5)$$

using the Abel inversion

$$\Delta n_m(r) = n_m(r) - n_{clad}$$

$$= -\frac{\lambda}{\pi D} \int_r^R \frac{dS(y)}{dy} \frac{dy}{(y^2 - r^2)^{1/2}}. \qquad (9.6)$$

Boggs *et al* (1979) and Presby *et al* (1979) described an automated transverse interferometric method and deduced the index profile of graded-index fibres. In transverse illumination the light is refracted primarily by the outer cladding boundary, the additional phaseshift and refraction contributions of the core having only a minor effect. This makes it advantageous to remove the effect of the outer cladding by immersing the fibre in matching oil with the same refractive index n_{clad}. Each light ray passes through the regions of varying refractive index and the total optical path length is expressed as an integral $OPL = \int_{S_1}^{S_2} n(s) \, ds$. The integration variable S is the path length measured along the light ray. The fibre is immersed in a drop of the matching liquid in which the microscope objective is dipped. This constitutes the arrangement in one arm of the interference microscope, while the other arm contains only a drop of the matching oil of refractive index $n_L = n_{clad}$

$$n(r) - n_L = - \frac{\lambda}{\pi D} \int_r^R \frac{dS(y)}{dy} \frac{dy}{(y^2 - r^2)^{1/2}}. \qquad (9.7)$$

The refractive index distribution of the fibre can thus be computed from fringe shift observations in transverse illumination obtained with the interference microscope in a single-pass arrangement. Processing of the data, the fringe shift, requires that a numerical differentiation is performed first, followed by the numerical integration indicated by equation (9.7). Boggs *et al* (1979) in their automatic single-pass profiling measurements inserted a fibre of length one centimetre into a matching index oil in the sample arm of a Leitz dual-beam, single-pass, transmission interference microscope. A similar thickness of matching oil is placed in the reference beam. For the wavelength of monochromatic light used, $\lambda = 0.9 \ \mu m$, $n_L = 1.457 \pm 0.0005$, and heating of the oil to achieve the matching condition was not found necessary. The sequential measurement procedure is similar to that of Wonsiewicz *et al* (1976) shown in figure 9.5 and Presby *et al* (1978), and involves video detection and digitisation of the interference fringes under computer control. The output field of the microscope is detected with an infrared-enhanced silicon target vidicon. The video signal is sent to a video digitiser that has the capability of addressing and encoding discrete picture elements in the television frame. The digitiser resolves 480 picture elements on the y axis and 512 elements on the x axis. The x and y position data inputs are provided by a 16-bit duplex input/output (I/O) interface of a Hewlett-Packard 9825A computer. Encoding is 8 bits or 256 gray levels and a digitised video is received by the calculator as 8 bits, parallel binary. The digitiser also incorporates a video output display which permits observation of the scene being processed and monitoring of the encoding on the same screen. The objective of the encoding process is to accumulate intensity versus position data so that the displacement of a fringe from its cladding level can be determined accurately as a function of radial position. Having determined the fringe displacement, the computer calculates Δn by a method termed the circular index method, a brief account of which follows shortly, and then plots the index profile along with coordinates and labelling on an xy plotter. The computer's program then determines a best-fit power law curve to the index profile, to deduce the value of α in the basic equation of the index profile of a graded-index fibre.

As explained, the output field of the microscope is automatically processed with a video-digitised computer-controlled system, the index profile is determined by the solution of an integral equation. The resulting profiles are reproducible to about 1% and can be determined within a few minutes of fabrication, conforming to the required optimum distribution. Presby *et al* (1979) reported that the explicit solution of the integral equation (9.7) showed how $n(r)$ at any point r in the fibre core could be obtained from the fringe shift $S(y)$ by differentiation and integration. Since the fringe shift was known only at certain discrete points, numerical techniques for approximating the derivative as well as the integral had to be used. The accuracy of the resulting refractive profile depends on the accuracy of measuring $S(y)$, on the density of points at which $S(y)$ was measured and on the sophistication of the methods used for numerical evaluation.

In the method of analysis performed by Boggs *et al* (1978), termed the circular index method, they assumed that rays pass the fibre core without deflection and their phases are retarded according to the length of their optical paths. In addition they assumed that the fibre core consists of a large number of concentric rings each with a constant refractive index. They evaluated the index step-by-step beginning with the periphery and proceeding towards the centre as it is possible to determine the index value of each ring if the index values of all preceeding rings are already known.

It is to be noted that it is only in differential interferometry, for example when using a Mach–Zehnder interferometer with a shearing device, as explained in Chapter 3, that the index distribution is given more directly by equation (9.7) as

$$\Delta n(r) = -\frac{\lambda}{\pi Ds} \int_r^R S(y) \frac{dy}{(y^2 - r^2)^{1/2}} \tag{9.8}$$

s being the lateral shift between the two beams in which computation of the derivative is no longer necessary.

In conclusion, the set-up, comprising an interference microscope, a vidicon camera, a video digitiser, a display monitor, a programmable calculator and a plotter, is suitable for performing automatic index profiling of a fibre when applying the interferometric slab method, the transverse interference method and single- or double-pass localised two-beam or multiple-beam fringes. In each case a microinterferogram is secured and recorded or viewed through a video camera, followed by a video digitiser, and displayed on a monitor. Clearly, the difference between the interference systems is in the optical path difference between interfering beams and consequently in the relations between the refractive index and the fringe shift, requiring appropriate programs for deducing the index profile. In the case of an interferogram resulting from the interferometric slab method, a linear relation exists between $\Delta n(r)$ and the fringe shift Δy in equation (9.4), while in the case of transverse interference systems, the relation between $\Delta n(r)$ and $S(y)$ is governed by an integral equation (9.7), an Abels' integral equation. Its solution shows that $\Delta n(r)$ at any point in the fibre core can be obtained from the fringe shift $S(y)$ by differentiation followed by integration. In differential interferometry the index distribution is given more directly by an integral equation in which computation of the derivative is no longer needed.

References

Billingsley F D 1971 Digitization and storage of the image in *Advances in Optical and Electron Microscopy* ed. R Barer and V E Cosslett vol. 4 (London: Academic) pp127–70

Boggs L M, Presby H M and Marcuse D 1979 *Bell Syst. Tech. J.* **58** 867

Eccles M J, McCarthy B D and Rosen D 1976a *J. Microsc.* **106** 33

—— 1976b *J. Microsc.* **106** 43

Hopkins H H 1943 *Proc. Phys. Soc.* **55** 116

Marcuse D and Presby H M 1980 *Proc. IEEE* **68** 676

Presby H M, Marcuse D and Astle H W 1978 *Appl. Opt.* **17** 2209

Presby H M, Marcuse D, Astle H W and Boggs L M 1979 *Bell Syst. Tech. J.* **58** 883

Rosen D 1984 Instruments for optical microscope image analysis in *Advances in Optical and Electron Microscopy* ed. R Barer and V E Cosslett vol. 4 (London: Academic) pp323–45

Wonsiewicz B C, French W G, Lazay P D and Simpson J R 1976 *Appl. Opt.* **15** 1048

Author Index

Subject Index